RESEARCH IN FLUVIAL GEOMORPHOLOGY

Proceedings of the 5th GUELPH SYMPOSIUM ON GEOMORPHOLOGY, 1977

WITHDRAWN

Edited by

R. DAVIDSON—ARNOTT

W. NICKLING

Published by

GEO ABSTRACTS LTD.,
University of East Anglia,
Norwich, NR4 7TJ, England

in association with

GEOMORPHOLOGY SYMPOSIUM,
Department of Geography,
University of Guelph,
Guelph, Ontario, Canada, N1G 2W1

Geographical Publication No. 5.

ISBN 0 86094 013 6

Information about the Guelph series of Symposia
in Geomorphology may be obtained from

'Geomorphology Symposium',
Department of Geography,
University of Guelph,
GUELPH, Ontario, Canada.

*The cover photograph of the Yukon river is by
W. Nickling*

Copies of these volumes (for a complete list see
page 4) may be obtained from the sole publishers:

Geo Abstracts,
University of East Anglia,
NORWICH, NR4 7TJ, England.

CONTENTS

3

PREVIOUS BOOKS IN THIS SERIES

No.1.   Research Methods in Geomorphology,
        1969,   140pp

No.2.   Research Methods in Pleistocene
        Geomorphology, 1971,   285pp
        0 902246 17 8

No.3.   Research in Polar and Alpine
        Geomorphology, 1973,   206pp
        0 902246 22 4

No.4.   Mass Wasting 1976,   202pp
        0 902246 58 5

23

...UTH, INC.

...ge Stre... ...ensboro, N. C. 27420

...by: | Phone No.: | Date Sent to Binder

ALL INSTRUCTIONS ON BINDING TICKET WILL BE FOLLOWED EXPLICITLY

LETTER SPINE EXACTLY AS FOLLOWS:	INSTRUCTIONS TO BINDER
Title:	___ Bind as is (with covers & ads)
	___ Remove front covers
	___ Remove back covers
	___ Remove ads (front & back)
	___ Remove all ads (extra charge)
	___ Bind title page/contents in front
	___ Bind index in front
	___ Bind index in back
Vol.:	___ Hand sew if necessary (extra charge)
	_____ Cover Color
	Letter in: ___gold; ___black;
Year	___white
	LIBRARY BOOKS:
	___Decorated covers; ___plain covers;
	___picture covers (extra charge)
Call No.:	

No trim

> **SUPER-FLEX** (economy binding) Uniform height, white lettering. Covers & ads bound in. Cover color for **periodicals only** _____.
> Books & Paperbacks — cloth colors random selected by binder; binder's choice of black or white lettering.
> Paperbacks: ___Mount front cover; ___Bind in covers; ___Discard covers.

76374 39

28

Special Instructions:

Send two copies of binding slip with volume; retain one copy for your files. If item returned for correction because of binder's error, original binding slip **must be** returned with volume.

FOREWORD

I am pleased to introduce this volume of the proceedings of the Guelph Symposium on Geomorphology, now the fifth in the biennial series. The Department of Geography is proud of its tradition of the regular meetings devoted to the study of geomorphological processes. Professor Eiju Yatsu, who returned to Japan in 1976, deserves special recognition for his role in the initiation of the series and in the establishment of geomorphology at Guelph.

All symposia represent a great deal of effort by the organisers and the contributors. Much of the early planning and logistical work was carried out by Barry Fahey. Particular thanks are due to Robin Davidson-Arnott and William Nickling who planned the programme during their first year in the department. Many people contributed to the success of the meeting and the preparation of the proceedings. We are grateful to the Office of Continuing Education, University of Guelph, for their work on the logistics of the symposium. The editors and the department are grateful to the staff technicians, Brian Reynolds and Mario Finoro; to the secretaries Lynn Cawthra and Becky Nixon, and to the cartographers, Margo Adamson and Pat Banister.

We acknowledge with gratitude the financial support of the National Research Council of Canada. Our thanks are also due to the speakers, session chairmen and participants who made the symposium a success.

GERALD T. BLOOMFIELD
Chairman
Department of Geography
University of Guelph

Part 1 FLUVIAL PROCESSES

AND SEDIMENTATION

1 MECHANISMS AND HYDRODYNAMIC FACTORS

OF SEDIMENT TRANSPORT IN ALLUVIAL STREAMS

*Roscoe G. Jackson II

ABSTRACT

*The analysis of sediment transport and open-channel hyd-
raulics can provide substantial information on the genesis
and behavior of fluvial features of geological pertinence.
The following topics of investigation require techniques
that involve multidimensional considerations of hydro-
dynamics and channel geometry and of detailed characteristics
of grain-size distributions of alluvial sediment: 1) scales
of physical processes of sediment transport (including flow
structures and bedforms), 2) mechanisms and rates of sediment transport in streams, and 3) grain-size distribution
of bed material. Patterns of flow velocity, bedforms,
channel geometry, and bed-material size from a meandering
river illustrate the failure of one-dimensional methods to
appraise adequately the preceding three topics. Satisfactory
explanation of these and other complex features of alluvial
flows requires utilization of the sophisticated tools of
flow and particle visualization, measurement of directional
components of hydraulic parameters (especially velocity),
adequate delineation of four-dimensional flow fields, and
computational fluid dynamics.*

INTRODUCTION

Sediment transport in alluvial channels has engaged the
attention of engineers and other physical scientists for
well over a century (Graf 1971, pp 1-24). Hydraulic
properties of open-channel flows have been examined for a
much longer time. Despite the long and honored history of
both fields of investigation, it has been only in the past
30 years or so that sedimentologists and fluvial geomorphologists have begun to examine the geological aspects of
rivers with the tools of the hydraulic engineer (eg.Sundborg
1956; Leopold, Wolman & Miller 1964; Allen 1970a). Because
the physical processes of sediment transport obviously play
a vital role in the genesis of many features of great
geological interest - such as primary sedimentary structures,
channel geometry, and grain-size distribution - it has been
only natural for the geologist to pursue the engineering
disciplines of sediment transport and open-channel hydraulics. Recent treatises by Leopold *et al* (1964), Allen

---

*Department of Geological Sciences, Northwestern University,
Evanston, Illinois 60201, USA

1. Sediment transport in alluvial streams

(1968, 1970a), and Church & Gilbert (1975), show the deep insight into fluvial geomorphology attainable from formal considerations of these subjects.

In spite of the eager acceptance of the significance of hydraulic engineering for fluvial geomorphology, it has become increasingly apparent during the past several years that many of the more traditional methods of hydraulic engineering have reached their limit of usefulness for the rational explanation of many fluvial features. One striking example which comes immediately to mind is the flow-regime classification of bedforms (Simons & Richardson 1961; Simons, Richardson & Nordin 1965). This approach employs simple concepts of one-dimensional steady-flow hydraulics for the empirical prediction of stability fields of flow-regime bedforms monitored under controlled laboratory conditions. One-dimensional parameters of open-channel hydraulics such as the Chezy coefficient, water-surface slope, and channel-mean size of bed material, are incorporated into this methodology to predict effectively the particular bedform type that occurs under a set of ideally simple laboratory conditions of flow, channel geometry, and sediment size. Geologists rapidly realized the applicability of this empirical approach to the explanation of common fluvial bedforms (Allen 1963; Harms & Fahnestock 1965; among others) and to the formation of primary sedimentary structures (several papers in Middleton 1965). Later studies, however, have demonstrated that the flow regime approach cannot explain satisfactorily many aspects of fluvial bedforms (Allen 1973; Costello 1974; Yalin 1975; Jackson 1976a). The difficulty arises from the inherent complexity of natural alluvial channels, in which the flow is neither steady nor one-dimensional, the channel never straight, and the sediment rarely as well sorted as in the laboratory. A completely rational explanation of the geomorphological parameters of most natural streams requires a correspondingly more complex treatment of the governing physical processes.

The present report describes several examples from fluvial channels in which one-dimensional hydraulics and simple parameters of grain size and channel geometry fail to determine adequately the observed geological features. In some instances use of more sophisticated principles of sedimentation can yield fuller explanations. In other cases application of such principles is clearly called for. Most of the examples to be cited come from the writer's research experience, but pertinent studies by others will be examined. The topics are restricted to unidirectional open-channel flow of small sediment concentrations. This paper is not a comprehensive review of the state of the art in sedimentation and flow in alluvial channels. Treatises that do furnish such appraisals include Graf (1971), Bogárdi (1972), Yalin (1972), Raudkivi (1976), and Vanoni (1976).

1. Sediment transport in alluvial streams

Table 1 Notation

A, B	dimensionless constants (Table 2)
d	mean grain size of bed material
$O(x)$	of the order of magnitude of the parameter x (Table 4)
Q	stream discharge (m^3/s)
Q_s	sediment discharge per unit channel width ($gcm^{-1}s^{-1}$)
T	periodicity of major events of sediment transport (Table 4)
T_E	Eulerian integral time scale; defined to be $\int_0^\infty R(\tau)d\tau$, where $R(\tau)$ is the one-dimensional autocorrelation function with time lag τ.
T_t	time scale of largest turbulent eddies of flow field (Table 4)
$\overline{T}_1$	mean periodicity of bursts at a point in the outer zone of the turbulent boundary layer
$\overline{T}_2$	mean periodicity of boils at a point on the water surface in a unidirectional open-channel flow
U	mean velocity of channel cross section
$\overline{U}$	local depth-averaged velocity
U_∞	free-stream velocity
u_*	shear velocity
u_{100}	local velocity 100 cm above bed
w	local width of channel at sediment-transporting flows (Table 4)
X	average sediment concentration; equals $Q_s/\rho U\delta$, where ρ is fluid density
δ	local thickness of boundary layer; equals local flow depth in nearly all alluvial channel flows
λ	wavelength
ν	kinematic viscosity of fluid
σ	standard deviation

--- Superscripts ---

x^+	dimensionless rendering of parameter x with inner variables; equals u_*x/ν
x^*	dimensionless rendering of parameter x with outer variables; equals $U_\infty x/\delta$

11

1.   Sediment transport in alluvial streams

THE PROBLEM

Sediment transport in low-concentration open-channel flows
is a two-phase flow, in which sedimentary particles constituting the solid phase are propelled by the shearing
flow of the liquid phase. Exact constitutive equations of
conservation of mass, momentum, and energy, can be written
for each phase (Soo 1967, pp 249-255; Yen 1973). The two
systems of coupled time-dependent second-order nonlinear
partial differential equations contain more unknown variables than equations, so exact solutions are not feasible.
The presence of turbulence, a virtually ubiquitous phenomenon in natural alluvial flows, introduces additional unknown parameters in the form of covariance terms $u'v'$.

The mathematical intractability of the governing dynamical equations renders the theoretical investigation of
sediment transport and fluid flow enormously more difficult
than other major fields of natural science in which the
governing equations can be solved analytically. Attempts
to simplify the governing equations have revolved around
phenomenological theories of simple relationships between
groups of unknown parameters and around assumptions of constant values for certain variables. By sufficient use of
both practices the dynamical equations can be simplified to
the extent that exact analytical solutions are obtainable
for any flow. However, the requisite assumptions behind
these simplifications usually are not met in natural
alluvial flows. For example, all the one-dimensional flow
equations require the suppression of cross flow and vertical flow and the existence of straight channels of regular
cross section (Yen 1973). No natural channels obey these
drastic assumptions; and it is not surprising, consequently,
that the one-dimensional equations often fail to explain
adequately the hydraulic behavior of natural streams.

During the past several years the science of fluid
flow has seen the advent of several promising approaches.
The potential applications of these new methodologies to
sedimentation in alluvial channels will be pointed out at
appropriate places in the following text.

PHYSICAL SCALES OF SEDIMENT TRANSPORT

Geomorphologists have long realized that sediment transport
in rivers takes place on a variety of time and length
scales. A corresponding variety of physical processes is
associated with these transport scales. However, only
recently have explicit scaling relationships been proposed
in order to explain some broader aspects of the genesis of
bedforms (Allen 1966; Church & Gilbert 1975, pp 53-61;
Jackson 1975a).   This section summarizes a unified treatment of transport scales and points out the relevance to
bedforms and physical processes of sediment transport.

The Turbulent Boundary Layer

The concept of a boundary layer is one essential component
of any treatment of physical scales of sediment transport.
The pertinence of boundary layers to sediment transport and

fluid flow has been appreciated by hydraulic engineers
(eg. Yalin 1972, pp 20-52) and by fluvial geomorphologists
(eg. Sundborg 1956, pp 138-146, 209-211;   Briggs &
Middleton 1965;   Allen 1970a, pp 36-49).   The main geological applications of boundary-layer principles have dealt
with grain roughness of sediment beds (eg. Allen 1964;
Allen 1970a, pp 75-81;   Moss 1972), form roughness and lee-
side flow characteristics of sediment ripples such as dunes
(Allen 1969;   Smith 1969;   Smith & McLean 1977a), settling
behavior of solid particles in a liquid (Rubey 1933;   Brush,
Ho & Yen 1964;   Murray 1970), and sorting of bed material
(Moss 1972).   Nearly all of these investigations dealt with
the classical topics of boundary layers such as the
definition of regions of the turbulent bounday layer in
terms of velocity profiles (Table 2) and the description of
bed roughness by the velocity profile.   These methods are
one dimensional and involve steady flow.   However, recent
experimental and computational developments in boundary-
layer investigations have provided a renewed appreciation
among fluid dynamicists of the inherent three dimension-
ality of turbulent boundary layers, even in seemingly simple
flows.   The following two sections present a general
analysis of two-phase flow in terms of these new methodologies.   Some obvious implications to bedforms and alluvial
transport will be examined.

Turbulent Structure

There presently is widespread agreement among fluid dynamicists that a significant degree of organization exists in
the turbulent fluctuations of velocity in turbulent boundary
layers (Laufer 1975).   A continuing series of flow-visualization experiments has gradually brought forth a fairly consistent picture of the detailed fluid motions comprising
the organized components of shear turbulence.   This basic
turbulent structure is summarized in Table 3, where it is
seen to be divided into two fundamental components.   The
inner zone of the turbulent boundary layer, including roughly
the sublayer and transition regions (Table 2) of the classical structure, displays spanwise alternations of comparatively
low-speed fluid and relatively high-speed fluid that are
elongate in the downstream direction.   These "wall streaks"
exert a marked influence upon near-bed conditions of
sedimentation (Table 3) and upon the existence of the class
of smallest bedforms, microforms (Jackson 1975a, 1976b).
The existence and spacing of wall streaks depend upon
variables of the wall (ie. bed) such as roughness, kinematic
viscosity, and wall shear stress (Table 3).   This dependence
in general is true for the inner zone of the turbulent
boundary layer (Tables 2 and 3).

On the other hand, the outer zone of the turbulent
boundary layer is much less dependent upon wall conditions,
and its essential properties tend to scale instead with
variables of the gross flow.   The turbulent structure of the
outer zone displays a quasi-organized sequence of fluid
motions collectively referred to as bursting (Offen & Kline
1974, 1975;   Nychas, Hershey & Brodkey 1973).   Individual

1. Sediment transport in alluvial streams

Region	Turbulent length scale	Velocity scale	Velocity law	y^+ range	Typical distance from wall #
viscous sublayer	$10\nu/u_*$	u_*	$u/u_* = y^+$	$0 < y^+ < 5$	$y < 0.25$ mm
transition	y	u_*	+	$5 < y^+ < 70$	0.25 mm $< y < 3.5$ mm
logarithmic	y	u_*	$u/u_* = A \ln(y^+) + B$	$70 < y^+ < 0.15\delta$	$y > 3.5$ mm
outer (wake)	δ	U_∞	$(U_\infty - u)/u_* = \text{fct}\ (y/\delta)$	$0.15\delta < y^+ < \delta$	
interface (superlayer)	$10\nu/U_\infty$	U_∞	U_∞ = constant	$y^+ > \delta$	

Note: Table adapted from Kovasznay (1967, Table 1) and from Teleki (1972, Fig. 7).

For $\nu = 0.01$ cm^2/s and $u_* = 2$ cm/s.

+ Transitional between laws for sublayer and logarithmic regions.

Table 2 Classical structure of turbulent boundary layers in steady incompressible single-phase flows over smooth walls.

	Inner zone	Outer zone
	----- Basic turbulent structure -----	
Extent	$y^+ \lesssim 50$	$y^+ \gtrsim 50$
Prevalent fluid motions	Wall streaks in viscous sub-layer; lift-up stage of burst cycle	Oscillatory growth and breakup stages of burst cycle
Critical fluid-dynamic variables	u_* d ν (inner variables)	U_∞ δ (outer variables)
Scaling relations	$\lambda_s^+ \simeq 100$	$\bar{T}_1^* \simeq 5$ $\gamma_1 \simeq 0.3$
Relative contributions of bursts and sweeps to Reynolds stress	Sweeps predominate	Bursts predominate for $y^+ \gtrsim 100$
Effects of bed roughness	Grain roughness disrupts wall streaks when $d^+ \gtrsim 10$	Form roughness (esp. ripples) localizes generation of bursts to areas of adverse pressure gradients. Grain roughness intensifies bursting
	----- Relation to other turbulent phenomena -----	
Boils and kolks	Lift-up initiates kolk Kolks preferentially form in areas of adverse pressure gradients	Kolk = Oscillatory growth stage Boil = Late oscillatory growth + breakup stages of burst cycle
Scaling relations of boils		$\bar{T}_2^* \simeq 7.5$ $\varepsilon/\delta \simeq 0.4$
Eulerian integral scale (T_E)		$T_E^* = \bar{T}_{1B}^* \simeq 1.75$
	----- Relation to bedforms and sediment transport -----	
Bedforms (Jackson, 1975a)	Microforms governed by turbulent structure of inner zone	Mesoforms governed by fluid-dynamic regime of outer zone
Scaling relations of bedforms	Current lineations: $\lambda^+ \simeq 100$ Small-scale ripples: $d^+ < 13$ † Small-scale ripples: $\lambda/d \simeq 1000$	Dunelike large-scale ripples (Dunes): $\lambda \simeq 7\delta \simeq U_\infty \bar{T}_2$
Sediment dispersal	Wall streaks concentrate moving grains into low-speed wall streaks Grain roughness intensifies bursting to remove fine interstitial grains	Bursts produce vertical anisotropy in turbulence necessary to suspend sediment· Vigorous upward flow in bursts entrains more and coarser sediment than tractive forces alone.

† For small-scale ripples superimposed on dunes, $8 < d^+ < 40$.

Table 3 Summary of bursting phenomenon in turbulent boundary layers of geophysical flows. From Jackson (1976b, his Table 4).

1. Sediment transport in alluvial streams

bursts begin with the sudden uplift of a low-speed wall
streak, then grow rapidly (often in an oscillatory manner)
as they move away from the bed, and finally break up into
the chaotic motion of tiny eddies. The existence of
bursting in alluvial flows is convincingly shown by the
well-known phenomenon of "boils" on the water surface. The
specific association of boils and kolks to stages of the
bursting cycle is summarized in Figure 1, and the lines of
evidence are discussed in Jackson (1976b). Two pertinent
aspects of this argument are given below, because they
invoke additional supportive data which did not appear in
the original paper.

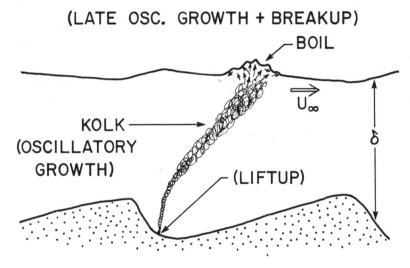

Figure 1.   Postulated relation of boils and kolks to
            stages of the Offen & Kline (1975) model
            of the bursting process. The relation
            also exists over flat beds. The fluid
            motions are in a time sequence of liftup,
            oscillatory growth, and breakup. From
            Jackson (1976b, his Figure 6).

The laboratory studies of bursting demonstrate a virtual
independence of this turbulent structure from wall conditions (eg. Rao, Narasimha & Badri Narayanan 1971; Grass
1971; Laufer 1975). Instead the characteristic dimensions
of bursts appear to scale with variables of the outer flow
(Table 3). Several recent investigations have demonstrated
the general validity of the following scaling relationship
(notation in Table 1)

$$U_\infty \bar{T}_1 / \simeq 5$$

in a variety of pressure gradients and in compressible
flows (air) as well as in incompressible flows (water).
Jackson (1976b, his Figure 7) found that boil periodicity $\bar{T}_2$
scales in a similar manner as burst periodicity. Figure
2 presents additional data to support further this contention.
Noteworthy in Figure 2 are the flat-bed data, which indicate

16

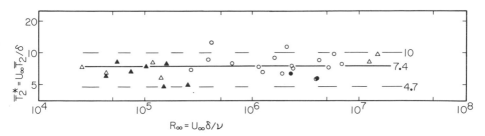

Figure 2.   Scaling of mean periodicity of boils ($\overline{T}_2$)
with the outer variables U_∞ and δ.
Circles denote data in Jackson (1976b, his
Figure 7 and Table 1).   Triangles refer to
data collected during 1976 from a variety
of natural streams.   Solid symbols denote
flat beds;   open symbols denote small-
scale ripples or dunes.   The heavy
horizontal line is the grand mean of 41
observations, 10 of which are not plotted
(see Jackson 1976b, his Table 1 for
explanation).   The two dashed lines define
a bandwidth of 2.1, which contains all but
two of the 41 total points.

that the scaling relation does not depend upon the substantial form roughness of large bedforms, although the
dimensionless periodicity $\overline{T}_2$* may be slightly lower for flat
beds than for rippled beds (6.5 for 9 flat-bed measurements
versus 7.7 for 32 measurements over rippled beds).

Jackson (1976b) argued that a second parameter of periodicity, the Eulerian integral time scale (T_E), likely was
equal to the mean duration of individual bursts at a point
in the flow.   The data he presented to support this hypothesis came from only one source, though a great range of
flow conditions and bed roughnesses was included.   Figure 3
presents a slighly modified scaling (using $\overline{U}$ instead of U_∞)
to incorporate the measurements of Grinvald (1965).   These
new data fall within the same narrow range of $\overline{U}T_E/\delta$ as
McQuivey's (1973) results, which the original diagram
employed.   The general constancy of this dimensionless
rendering of T_E with respect to flow Reynolds number is
particularly remarkable in view of the large size of some of
the block and rock roughnesses, some of which exceeded $\delta/2$
(Jackson 1976b, his Table 2 and p 547).

Two possible implications of the bursting phenomenon
to sedimentation can be envisioned (Table 3).   Firstly,
there is strong circumstantial evidence that bursting produces dunes of the flow-regime classification.   Yalin (1976)
has arrived independently at this conclusion.   The second
implication maintains that the astonishingly powerful upward flow velocities and large local pressure gradients in
bursts play a major role in the entrainment of sediment from
the bed and in the suspension of sediment.   If these hypotheses are valid - and they certainly bear much additional

1. Sediment transport in alluvial streams

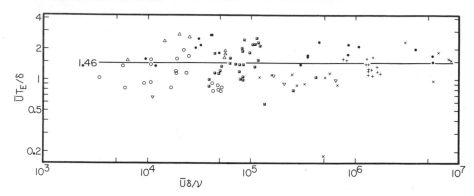

Figure 3. Scaling of Eulerian integral time scale (T_E) with outer variables $\bar{U}$ and δ, emphasizing bed roughness. Heavy horizontal line is arithmetic mean of all 107 plotted points. Data denoted by + come from Grinvald (1965, his Table 2) and were obtained from the Turunchuk river on 16 August 1962. The river bed at the time of these measurements showed a "wavy form" (Grinvald 1965, p 91), which presumably meant the presence of small-scale ripples or dunes. Kinematic viscosity for the Grinvald data is assumed to be that of pure water at 21°C, the mean August temperature in nearby Odessa. The remaining data come from McQuivey (1973) and are depicted in a similar diagram in Jackson (1976b, his Figure 13). Δ, smooth rigid bed; o, shot or rock roughness; o, lower-regime or upper-regime flat bed (0.1 mm < d < 0.4 mm); X, small-scale ripples or dunes (excluding dune troughs) or transition (0.2 mm < d < 0.5 mm); block roughness; ∇, averaged values of trough and crest of a dune. Data for the latter symbol are distinguished in order to reduce substantial turbulence differences caused by the local form roughness of large dunes in shallow flume flows. These data were not differentiated in Jackson (1976b, his Figure 13 and Table 2).

examination theoretically and in the laboratory - then many of the conventional expressions for thresholds of sediment movement and sediment suspension and for rates of sediment transport may need to be modified to reflect more directly the turbulent contribution. Indeed, Kalinske (1943) long ago argued that the turbulent fluctuations in rivers are much more important than mean-flow parameters in predicting the conditions under which sediment is entrained and suspended. The writer's ongoing study of this phenomenon indicates that the vertical anisotropy in turbulence arising from bursting (Table 3) does indeed exist in natural turbulent boundary layers and is of sufficient magnitude to be an important factor in suspension.

The preceding considerations of one type of flow structure, that arising from shear turbulence, in turbulent boundary layers lead to analysis of topics not feasible in the conventional one-dimensional hydraulics. The proposed dependence of dunes upon bursting and of small-scale current ripples and current lineation upon wall streaks requires careful examination in the laboratory in order to evaluate further this proposal. Instead of measurement of gross-flow parameters, near-bed conditions and three-dimensional fluid motions will have to be measured in the three spatial dimensions and the fourth dimension of time. Although statistical studies of ripple wavelength have started in this direction (Jain & Kennedy 1971, 1974), there clearly is a definite need for visualization experiments and turbulence measurements in alluvial flows over mobile beds.

Flow structures

The proliferation of flow-visualization and experiments by fluid dynamicists (eg. reviews by Laufer (1975) and Roshko (1976)) suggests that all turbulent shear flows possess one or more inherent flow structures. Each flow structure consists of a secondary motion superimposed upon the predominant unidirectional mean flow and often is most strikingly manifest in vertical planes parallel or normal to the mean flow (Kline *et al*, 1967; Roshko 1976). Because fluid motions in such flow structures display large components normal to the confining boundaries of the flow, they may influence strongly sediment transport and the behavior of bedforms.

Table 4.   Scales of physical processes of
           sediment transport in open-channel flows.

	Macroscale	Mesoscale	Microscale
Time	$\gg T$	$< T$ or $O(T)$ $\gg T_t$	$> T_t$ $\ll T$
Wavelength	$O(10w)$ or $\gg w$	$O(5\delta)$	$O(50\nu/u_*)$
Governing factors	Major events of sediment transport	Outer zone of TBL; U,δ	Inner zone of TBL; u_*,d,ν
Effects on flow system	Entire system is modified	Local TBL is modified	Only flow near bed is modified
Bedforms	Macroforms	Mesoforms	Microforms

There appear to be three broad classes of flow structures. Each class corresponds to a class of bedforms and to a class of physical processes of sediment transport (Table 4). Microscale processes are summarized in Tables 3 and 4 and appear to be limited to the wall streaks of the turbulent boundary layer. Mesoscale phenomena, however, encompass a greater variety of flow structures and show definite associations with particular bedforms. Table 5

1.   Sediment transport in alluvial streams

gives a very tentative indication of the wavelength of meso-scale flow structures and of their correspondent bedforms. Transverse roll vortices listed in Table 5 have been observed in the planetary boundary layer but remain incompletely documented in aqueous flows.

Table 5.   Tentative relationship between wavelength of mesoscale flow structure and corresponding bedform.

Flow structure	Mesoform	Wavelength
Bursting	Dunes	$\lambda/\delta \simeq 7$
Longitudinal roll vortices	Longitudinal dunes Sand ribbons	$\lambda/\delta \simeq 2\text{-}6$
In-phase waves	In-phase bed waves	$\lambda/\delta \simeq 4$
Transverse roll vortices	??? perhaps transverse bars	λ/δ uncertain (may be about 30)

Each mesoscale flow structure contains a distinct fluid motion which appears in at least three cases to be closely associated with a single type of bedform, as suggested in Table 5.   Arguments to support these contentions and further ramifications are presented in a separate manuscript under preparation.   Table 6 presents one additional topic to be mentioned herewith.   This illustration suggests that a given flow regime contains a dominant flow structure that exerts an overriding influence on the bed response to the flow. There is now sufficient experimental documentation of the magnitudes of secondary flow in mesoscale flow structures to begin to compare the fluid motions in one structure to those in another.   One difficulty in this analysis is that the dominant flow structure can markedly affect the behavior of the subordinate flow structures, as evidenced by the localization of boils and bursts in an upper-flat-bed flow with strong longitudinal roll vortices (Coleman 1969, pp 205-206;   Müller 1976).

Table 6   Hypothetical sequence of dominant flow structures and bedforms in a unidirectional open-channel flow.

None prevalent	No movement of bed material
Microscale turbulent structure	small-scale current ripples
?? transverse roll vortices	Transverse bars
Bursts	Dunes
Longitudinal roll vortices	Longitudinal dunes or upper flat bed
In-phase waves	In-phase bed waves (antidunes)

Note: Sequence from top to bottom is for increasing velocity at constant depth (δ), with a bed of medium sand.

1. Sediment transport in alluvial streams

Macroscale processes transcend local hydraulic conditions (Table 4). The nature of macroforms has been elaborated by Jackson (1975a) and will not be reiterated. However, the writer has subsequently realized that certain large-scale flow structures which can scale with channel width (instead of depth, as in the case of mesoscale structures) exist in many alluvial flows. The spiral motion of fluid in a curved channel, for example, can be shown convincingly to produce point bars (Yen 1970; Onishi, Kennedy & Jain 1972). Another macroscale flow structure that has been thought to be significant in generating macro- forms is the helicoidal flow with a paired-vortex structure (Leliavsky 1955, pp 181-184; Einstein & Li 1958; Einstein & Shen 1964). (This secondary flow is conceptually distinct from the mesoscale longitudinal roll vortices discussed above, in that the wavelength of the former equals the channel width).

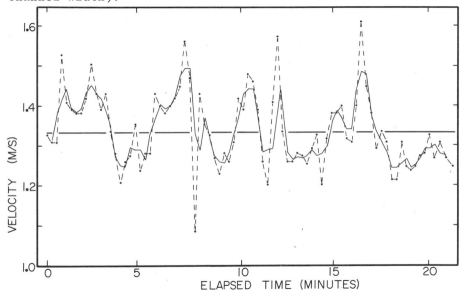

Figure 4   Temporal variation in velocity near Steveston Buoy S21 in the Fraser River at Vancouver, Canada. Velocity measured 6 m below water surface (5.2 m above bed) by a Price current meter on 27 May 1976. Each solid circle is a velocity computed from 30 revolutions of the bucket wheel of the Price meter. Segmented solid line is the running average of three consecutive velocities. Heavy dashed hori- zontal line denotes mean velocity of the measurement period. Time axis is not linear because velocities were plotted as equally spaced in time, although sampling time for each point varied inversely with the current speed.

1. Sediment transport in alluvial streams

One manifestation of a macroscale flow structure in
alluvial channels is the existence of long-period oscillations in flow velocity (Matthes 1947, p 256; USACE 1950).
Figure 4 shows one such measurement from the Fraser River.
The velocity is seen to vary regularly from 1.45-1.50 m/s
to 1.25 m/s with a period of perhaps 4 min. This periodicity
exceeds any turbulent time scale in having a dimensionless
value of $U\infty T/\delta$ of about 29. The regular oscillations about
the mean are not characteristic of the irregular fluctuations
due to shear turbulence (eg. Heathershaw 1974, 1976).
Whether or not this common type of long-period oscillation
is the manifestation of some mesoflow structure besides
bursting or of an as-yet-undisclosed macroscale structure
remains an open question. Preliminary considerations
indicate the likelihood that mesoscale longitudinal roll
vortices generate this type of velocity oscillation. Some
of the extremely long periodicities in the lower Mississippi
River (USACE 1950) must, in contrast, be due to macroscale
flow structures.

Implications to Scales of Sediment Transport

The preceding analysis of flow structures and bedforms
bears direct significance to physical scales of sediment
transport in open channels. Each flow structure contains
substantial fluid motions directed normal to the alluvial
bed and thus can produce the spatially non-uniform transfer
of sediment that is necessary for bedform genesis. Fluvial
processes of bed-material transport must therefore be
examined on three scales: 1) the smallest (in time and
space) being required for microforms and being restricted
to local flow conditions near the bed; 2) an intermediate
scale demands consideration of local flow conditions
throughout the turbulent boundary layer; and 3) the
largest scale requires analysis of flow structures that
scale with the largest dimension of the channel and whose
influence is felt only over times much greater than the
duration of individual events of sediment transport
(Table 4).

It is only fair to conclude this section with the
concession that at present the distinction between macroscale and mesoscale phenomena in alluvial flows is largely
one of conceptual idealization rather than one based upon
abundant documentation. There is an urgent need for
investigations of flow fields, transport events, and bedforms associated with mesoscale and macroscale processes.
Needless to say, these studies must go far beyond the one-
dimensional hydraulics employed so frequently in the
examination of alluvial sedimentation (eg. Allen 1970a-c;
Bridge 1975; Church & Gilbert 1975).

MECHANISMS AND RATES OF SEDIMENT TRANSPORT

The transport of bed material in low-concentration unidirectional flows appears to occur in three ways. The solid
particles may roll or slide along the bed (traction),travel
in intermittent contact with the bed (saltation, or intermittent suspension), or else be immersed in the flow without

any contact with the bed (true suspension). This classical
distinction has provided the framework for several theories
of bed-material transport and for practical formulas for
rates of bed material transport (Graf 1971, pp 123-242;
Yalin 1972, pp 111-203). Unfortunately for the sake of
simplicity, several factors can largely obscure this simple
subdivision of transport mechanisms and, in so doing,
present difficulties in the accurate computation of parameters of alluvial sediment transport.

Incipient Motion

Hydrodynamic conditions at the threshold of sediment movement are of obvious importance to the prediction of sediment
transport from parameters of the flow and sediment. The
traditional Shields' diagram (Graf 1971, p 96) comprises
the usual criterion for initiation of motion of bed material.
Several assumptions in the formulation of Shields' diagram
limit its effective application in many alluvial flows.
The most restrictive assumptions include steady uniform
flow, a flat bed, and sediment of uniform size and shape
(Yalin 1972, pp 79-81). The latter two assumptions appear
to be most severe, although a rippled bed can be treated by
analytical separation of the form roughness of the ripples
from the grain roughness of the bed proper (Smith & McLean
1977a, 1977b). The difficulty posed by poorly sorted sediment becomes most apparent in graveliferous streams that
show armoring (Gessler 1971). The armor consists of one
layer of the largest grains in the bed material and
protects the underlying sediment from entrainment. Armored
beds often show sediment transport only at the most
energetic flow conditions whereas unarmored portions of the
same stream bed can show appreciable rates of sediment
transport at much less intense hydraulic conditions.

The computation of flow parameters for Shields' diagram
normally involves one-dimensional hydraulic parameters such
as bed shear stress and mean grain size. However, more
complex methods are required for examination of many natural
alluvial beds which do not satisfy the assumptions mentioned
above. One additional complication arises from shear
turbulence. Local turbulent fluctuations in bed shear stress
can produce temporary conditions suitable for the entrainment of sediment grains locally, while mean-flow conditions
may fall well below those of the threshold curve for
initiation of motion. Grass (1970, his Figure 3) analyzed
this effect by determining the probability distributions of
instantaneous bed shear stress and for the instantaneous
critical shear stress associated with a given bed material.
He found that the two probability distributions showed
substantial overlap (ie. some grains could move) when the
mean bed shear stress was as little as 0.4 the mean critical
shear stress. Grass emphasized the vital roles of the
detailed packing arrangement of grains on the bed surface
and the turbulent fluctuations in bed shear stress in the
determination of the motion of an individual bed grain.
Much of the scatter in Shields' diagram, which exists even
under the ideally simple laboratory experiments, was
attributed to these two factors.

1. Sediment transport in alluvial streams

Saltation and Suspension

The role of turbulent structure in sediment suspension is summarized in Table 3. Although turbulence is generally thought to be a prerequisite for suspension of particles denser than the fluid in low-concentration flows, precise criteria for prediction of flow conditions under which suspension will take place are not available. It is commonly maintained that suspension of a solid particle will occur when shear velocity exceeds the settling velocity of the particle in still water (Inman 1949; Francis 1973; Middleton 1976). The physics of saltation remain poorly documented, but Francis (1973) has demonstrated the occurrence of saltation in laminar flow. Saltation is a ballistic phenomenon involving momentum transfer among colliding grains, but the analysis of the mechanics of saltation has rarely passed beyond estimation of the height and length of saltation jumps (Bagnold 1941; Einstein 1950; Owen 1964; Francis 1973).

Formulas for Rates of Sediment Transport

The widely recognized log-cycle scatter of predictions of bed-material transport rates from measured values scarcely bears mention. The inability of virtually all existing formulas to predict consistently transport rates to within a factor of two of measured values has been thoroughly documented (eg. White, Milli & Crabbe 1975; Yang 1976; Yang & Stall 1976), although new formulas purporting to offer superior prediction continue to appear in the engineering literature. The following considerations offer little hope for significant improvement of practical formulas of widespread applicability in rivers.

The existing transport formulas come largely from flume studies that employed uniformly sized and shaped sediment. Although some formulas, such as Einstein's (1950) and Bagnold's (1966), include expressions for a range of sediment sizes, the laboratory experiments and field measurements against which the formulas are compared invariably use uniform sediment. Most of the formulas emphasize the computation of transport rates for the entire cross section of the stream and consequently employ one-dimensional hydraulic parameters (eg. energy slope and average stream velocity) and assume steady flow. The presence of a wash load complicates matters further. Armored beds and nonuniform flow, especially spiral motion in curved channels, present other complexities which the existing formulas cannot handle readily.

In view of these difficulties it is not surprising that so few computations of sediment-transport rates have proceeded beyond prediction of total transport rates at a given cross section in a stream. The computation of sediment transport in most natural streams requires evaluation of cross-channel and down-channel variations in transport rates if the genesis of major channel features is to be explained in quantitative terms.

GRAIN-SIZE DISTRIBUTION OF BED MATERIAL

Among the many intriguing aspects of the statistical distribution of grain sizes in a sample of natural bed material is
the possibility that statistical characteristics of the
size distribution may reflect specific mechanisms of transport and may even be diagnostic of particular depositional
environments. A new approach in this direction was introduced by Visher (1969) and applied further by Visher and

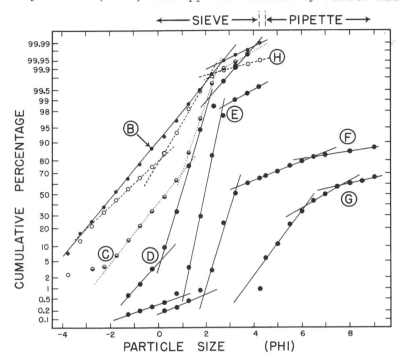

Figure 5   Log-probability plots of bed-material
            samples from the lower Wabash River near
            Grayville, Illinois. Circled letters in
            this diagram and in Figure 6 refer to the
            same samples. Sample HT2-2 (letter A)
            displays a log-probability plot like that
            of sample BT9-3 (letter H) and was
            omitted from this figure for clarity.
            Methods of grain-size analysis are
            explained in Figure 6.

his associates (Visher & Howard 1974; Freeman & Visher
1975) and by others (eg. Galloway 1976; Amaral & Pryor
1977; Saunderson 1977). This method entails a plot of the
cumulative distribution of grain sizes (determined from
sieving) of bed-material sample on log-probability paper,
as illustrated in Figure 5. Straight-line segments of the
distribution are assumed to correspond to truncated log-
Gaussian sub-populations and to represent specific modes of
transport. The grain size at truncation points between
adjacent segments and the percentage of the total distri-

bution contained within each line segment are deemed to be indicative of a particular depositional environment (Visher 1969, his Table 1). Figure 5 illustrates the fallibility of this latter assertion, for there appear segmented curves resembling those of turbidities (curve B), beach deposits (curve E), and tidal inlets (curve C). None of the log-probability plots in Figure 5 and few of the 200 constructed in this investigation (Jackson 1975c) resemble closely the ideal plots of fluvial deposits in Visher (1969). Reed, Le Fever & Moir (1975, p 1323) observed a similar lack of correspondence in their sieve analyses of fluvial sediments.

Further complications in the log-probability diagrams come from the production of spurious line segments in the regions of overlap of adjacent non-truncated log-Gaussian subpopulations. This matter is mentioned by Blatt, Middleton & Murray (1972, pp 40-41) and thoroughly examined by Reed et al (1975). The issue of whether the subpopulations are truncated Gaussian or non-truncated Gaussian remains undecided.

An alternative to log-probability plots of cumulative size distribution is the spline-frequency technique of Oser (1972) and van Andel (1973). This technique fits a cubic spline to the cumulative distribution, differentiates the spline curve to obtain the size frequency distribution, and then resolves that continuous curve into a sum of log-Gaussian sub-populations. The resolution can be accomplished by analog computer (Oser 1972) or by digital means. Figure 6 shows the size-frequency distributions of the samples of Figure 5. Whereas samples A and H show virtually identical log-probability curves in Figure 5, their size-frequency curves in Figure 6 are distinct.

The use of log-probability plots appears to be justified only for sediments of one dominant (non-truncated) subpopulation and with little overlap of adjacent subpopulations (see also Reed et al (1975, p 1322)). Samples D and E in Figure 5 are illustrative examples. More poorly sorted sediments with overlapping subpopulations would seem to be better analyzed by the spline-frequency technique, as witnessed by samples A-C and H in Figures 5 and 6. Samples with substantial pan fraction of unanalyzed sizes are best investigated with the log-probability technique (eg. sample G), because the size frequency method becomes unreliable at the extrema of the overall size distribution (Van Andel 1973, pp 439-440).

The possible correspondence of each subpopulation to a distinct mode of transport has been examined by Middleton (1976). He found it was possible to separate a traction subpopulation from an intermittent-suspension population in grain-size distributions from sand-bed plains rivers. The intermittent-suspension subpopulation included material that traveled in true suspension and in true saltation.

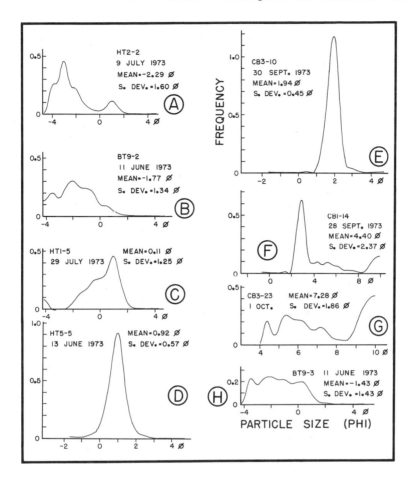

Figure 6   Typical size-frequency curves of bed material in the Grayville reach. Letters in sample labels are CB = cutbank section, B = Bozeman bend, H = Helm bend, and T = traverse. First number following the letters is the number of the cutbank or traverse; second number is the sample number or the station number, respectively. All traverse samples were subaqueous. Particle sizes coarser than 4.25 phi were measured by sieving at intervals of 0.5 phi. Sizes finer than 4.5 phi were determined by pipette analysis at intervals of 0.5 phi down to 9 phi; pan fraction recorded as 10 phi.

1. Sediment transport in alluvial streams

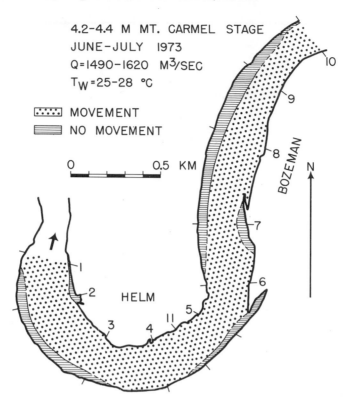

4.2-4.4 M MT. CARMEL STAGE
JUNE-JULY 1973
$Q = 1490-1620$ M^3/SEC
$T_W = 25-28$ °C

[⋯⋯] MOVEMENT
[≡≡] NO MOVEMENT

0   0.5 KM

Figure 7   Distribution of zones of movement and of no
movement of bed material around Bozeman and
Helm bends at near-bankfull flows. Movement criterion based upon mean size of bed
material and upon velocity profiles (to
determine u_*) at each site on traverses
1-9 and 11. Bed-material transport along
traverse 10 was inferred from the existence
of actively migrating dunes along the
entire traverse (Jackson 1975b, his
Figure 10).

COMPLEXITIES OF FLOW AND SEDIMENTATION
IN A NATURAL MEANDERING STREAM

The writer has completed an exhaustive study of sedimentation
in the meandering lower Wabash River (Jackson 1975c).
Hydraulic surveys revealed characteristic patterns of bedforms, bed topography, bed-material size, and flow velocity
around each of three bends monitored from low flows to
moderate floods (Jackson 1975b, 1976c). Large variations
in each of these fluvial features existed throughout a cross
section, downstream along each bend, among bends of disparate plan geometry, and with changes in stream discharge.
The patterns thus displayed a fully four-dimensional
behavior in space and time. The following remarks point

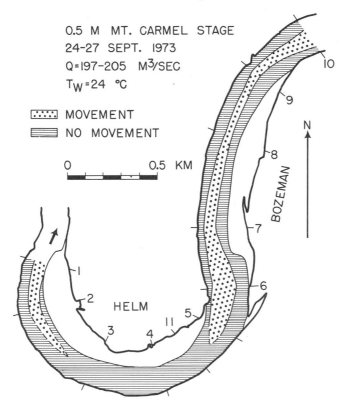

0.5 M MT. CARMEL STAGE
24-27 SEPT. 1973
Q=197-205 M³/SEC
T_W=24 °C

☷☷☷ MOVEMENT
≡≡≡ NO MOVEMENT

0   0.5 KM

Figure 8   Distribution of zones of movement and of
            no movement of bed material around
            Bozeman and Helm bends during seasonal-
            low flow. Crescentic white area on the
            inside of each bend is the subaerial
            portion of the point bar. Movement cri-
            terion based on the mean size of the
            bed material and on velocity profiles
            at each site on traverses 1-11.

out examples of this complex behavior which illustrate the
points raised in the preceding portions of the paper.

        In spite of the caveats mentioned earlier, Shields'
diagram provides useful, if approximate, indications of
threshold conditions for bed-material transport in many
alluvial flows. Figures 7 and 8 present computed zones of
bed-material movement and of no movement in two meander
bends of the lower Wabash River. The general picture is one
of prevalent transport at the high river stage and local
transport at much lower rates (see later Figures 10 and 11)
during the seasonal-low flow. The distribution of the zone
of sediment movement at the low flow is highly three
dimensional in that in a given traverse across the stream,
only a portion of the bed surface is undergoing transport
and that portion varies along the channel. An analogous

1. Sediment transport in alluvial streams

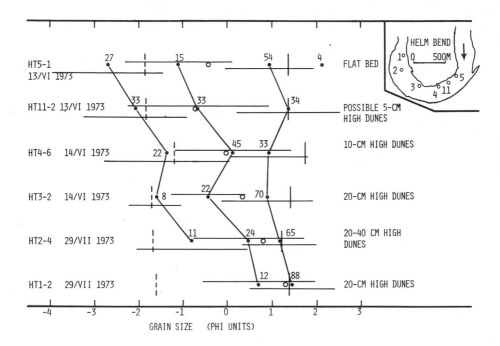

Figure 9 Grain-size composition and transport conditions
of 6 bed-material samples around Helm bend (inset).
Samples taken at a near-bankfull stage at Mt. Carmel
gaging station (Figure 7). Grain sizes determined by
sieving at intervals of 0.5 phi. Open circles are mean
grain sizes of samples; solid circles are modes of
log-Gaussian subpopulations determined from resolution
of size-frequency curves by a DuPont 310 Curve Resolver.
Horizontal lines show sorting of each subpopulation;
each line extends from -2σ to $+2\sigma$ of the respective subpopulations. This span includes 95% of the subpopulation.
Number beside each solid circle is the percentage of the
total sample consisting of the respective subpopulation.
Vertical dashed line for each sample is the threshold
size for grain motion on the bed, as determined from
Shields' diagram (Graf 1971, p 96); grain sizes to the
left of this line are predicted to be at rest under the
flow conditions at the time of sampling. Vertical solid
line for each sample is the threshold size for suspension
computed from a modified Bagnold criterion (Jackson 1975c,
pp 163-168); sediment to right of this line is predicted
to travel in suspension under the hydraulic conditions at
the time of sampling.

situation appears at flood flows, when bypassed portions
of the channel can show very little bedload transport
(Jackson 1975b, his Figures 5 and 11).

Grain-size distributions of bed material on the point
bar of the Helm bend are polymodal and contain overlapping
log-Gaussian subpopulations (Figure 9). Three subpopulations
can be identified in each sample in Figure 9 and traced
from sample to sample. The finest subpopulation changes
little in mode or standard deviation and shows a consistent
association with the calculated threshold size for suspension.
However, this threshold generally does not fall at either
extremity of the fine subpopulation. Observations of the
water surface during sampling revealed that much sediment
of the predominant size of this fine subpopulation (1-2
phi) appeared at the water surface at least temporarily in
boils (Jackson 1976b, p 554), which suggests at least an
intermittent suspension. This interpretation is at odds
with the fact that the suspension criterion shows little
sediment to have been deposited from true suspension
(Figure 9).

The Shields' criterion for incipient motion likewise
is not an accurate predictor of threshold conditions in
the samples of Figure 9. With the exception of sample
HT5-1 all the samples came from beds of actively migrating
bedforms. One possible explanation for this discrepancy
between prediction and observation in the upstreammost two
samples invokes the phenomenon of bed armoring, which
exists at lower flows and can be easily seen when the bed
is exposed at seasonal-low flow.

In sum, Figure 9 suggests that although samples along
the flow direction in a meander bend can show traceable
subpopulations, the correspondence of each subpopulation to
a particular mechanism of transport remains tenuous.

The large variation of local hydraulic conditions
with stream discharge in this river is illustrated further
in Figures 10-12. Figures 10 and 11 document the extreme
variability of transport rates with time and the disparate
response of different sections of the channel with changes
in stream discharge. This complex response appears
strikingly in Figure 12 for the entire cross section. The
general unreliability of a single exponential relation
between mean velocity and stream discharge for a stream is
apparent in Figure 12 and supports the similar, if less
spectacular, findings of Knighton (1975) and Bridge &
Jarvis (1976, their Figure 6). The (one-dimensional) concept of hydraulic geometry advanced by Leopold & Maddock
(1953) and extended by numerous subsequent studies does
not seem well suited for freely meandering streams.

Departures from ideal behavior in Figures 10-12 can
be rationalized qualitatively from a description of changes
in the flow zones defined for the Wabash meander bends
(Jackson 1975b, pp 1512-1515). The behavior of these flow
zones with changing stream discharge is a function of bed
topography and bend flow. Traverse 9 of the Bozeman bend
lay in a transitional (or "developing") flow zone, in which

1. Sediment transport in alluvial streams

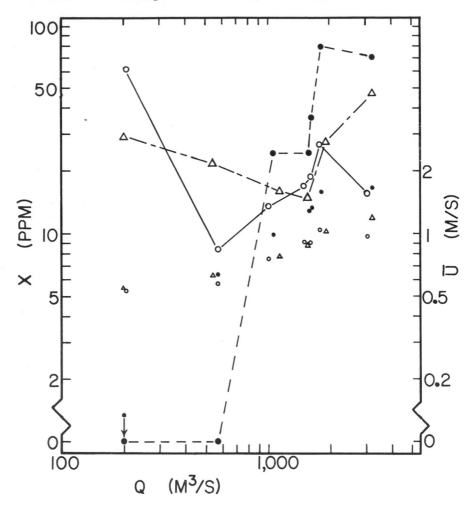

Figure 10   Variation of sediment concentration X and local
             depth-averaged velocity Ū with stream dis-
             charge Q at three positions in the lower
             Wabash River. Large symbols connected by lines
             denote values of X; small symbols denote values
             of Ū.   Triangles represent data from a position
             on Maier traverse 6 (Figure 14) about 145 m
             from the right (outer) bank; mean grain size of
             bed material ranged from 0.4 mm to 0.6 mm.
             Solid circles are measurements on Bozeman
             traverse 9 (Figure 7) about 55 m from the left
             (inner) bank; bed-material mean size ranged
             from 1.9 mm to 3.5 mm. Open circles denote
             data from Bozeman traverse 6 (Figure 7) about
             95 m from the left (inner) bank; bed-material
             mean size was about 0.7 mm. X calculated from
             total-load formula of Ackers & White (1973).
             Q is stream discharge at the Mt. Carmel
             gaging station.

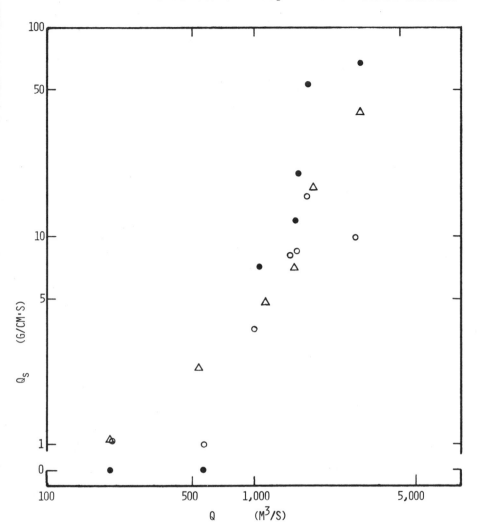

Figure 11   Variation of sediment discharge per unit
             channel width Q_s with stream discharge
             Q at three positions in the lower Wabash
             River. Symbols defined in Figure 10. Q_s
             computed from total-load formula of
             Ackers & White (1973).

current velocities increase monotonically from very low,
often near-zero, magnitudes at low flows to peak values at
flood flows (Jackson 1976c, his Figures 6 and 11). The
other two traverses of Figures 10-12 lay in zones of fully
developed bend flow, in which maximum velocities and bed
activity exist at discharges well below bankfull stages.

    A rigorous, quantitative explanation of these trends
will require examination of the three-dimensional spatial
properties of open-channel flow in these bends at different

1. Sediment transport in alluvial streams

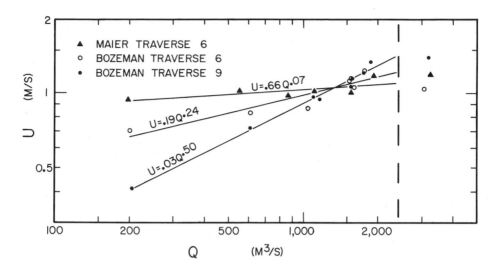

Figure 12   Variation of cross-sectional mean velocity
U with stream discharge Q for three
traverses across the lower Wabash River.
Traverse locations given in Figures 7 and
14. Q is the total stream discharge at
the Mt. Carmel gaging station. Vertical
dashed line is approximate discharge
(2,420 m³/s) at a 6-m Mt. Carmel stage,
above which overbank flow becomes significant at all three traverses. The three
regression lines include only those discharges less than this upper limit for
channel flow. U determined from measured
velocities across each traverse and from
U = Q/A, where A is the measured cross-
sectional area at the discharge Q. U for
the three flood discharges was determined
solely from velocity measurements.

stream discharges. In view of the mathematical intractability of the governing flow equations, an experimental
approach or numerical computation by approximation techniques seems called for. Figure 13 reveals the close physical
simulation of flow velocities that can be obtained in a
laboratory bend in which plan geometry, bed topography, and
stream discharge, are dynamically equivalent to these
features of the prototype natural bend. Use of the rapidly
improving tools of computational fluid dynamics for
numerical simulation appears justified by the encouraging
preliminary calculations of bend flows by Zelazny & Baker
(1975, their Figure 8), who used a finite element method,
and by Waldrop & Tatom (1976), who employed a fully three-
dimensional finite difference technique.

Figure 14 portrays the distribution of total bed shear
stress, which includes contributions from form roughnesses

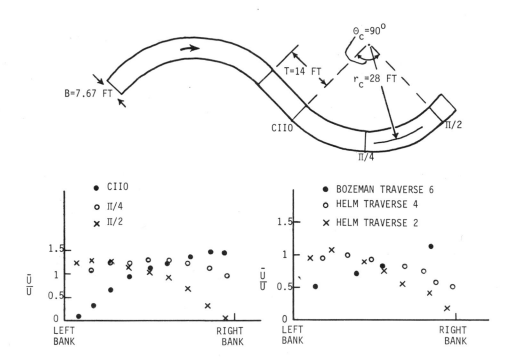

Figure 13   Measured velocities in cross sections of a laboratory bend and a natural bend. Upper panel
depicts geometry of laboratory channels used by
Yen (1970) and Onishi *et al* (1972). Lower left
panel gives distribution of velocity across
three cross sections of experimental run of Yen
(1970). Bed topography was a rigid replica of
the equilibrium topography of a natural sand
bed. Data come from Yen (1970, his Figure 5).
Figure 4.20 of Onishi *et al* (1972) illustrates
a velocity pattern virtually identical to that
of the lower left panel; Onishi *et al* used the
same laboratory bend with a mobile sand bed
instead of a rigid bend. Lower right panel shows
velocity distributions across three sections in
the lower Wabash River, whose locations correspond to those in the left panel, at near-bankfull
discharges equivalent to those of the left
panel. Figure 7 locates the three Wabash traverses, and the velocities plotted come from the
flow illustrated there (see also Jackson (1975b,
his Figure 3)). $\bar{U}$ = local depth-averaged
velocity; U = cross-sectional mean velocity.

1. Sediment transport in alluvial streams

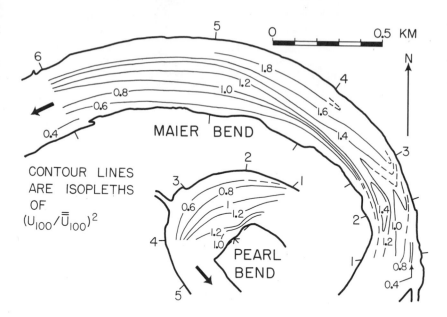

Figure 14   Distribution of shear-stress ratios
around Maier and Pearl bends at near-
bankfull flows. Ratio is local bed
shear stress divided by average shear
stress of entire bend (cf. Yen
(1970, his Figure 6a)). Shear stress
assumed proportional to u_{100}^2, following
the practice of Sternberg (1972).
u_{100} = velocity 1m above the bed.

of bedforms and the curved boundary, around two bends in
the lower Wabash River. The shear-stress pattern of
Maier bend is qualitatively very similar to the shear-stress
pattern reported by Yen (1970, his Figure 6a) over the
rigid-bed laboratory replica sketched in Figure 13. It too
is three dimensional in space.

CONCLUDING REMARKS

This selective appraisal of alluvial sedimentation has put
forth the following philosophy. The four-dimensional
properties of alluvial flows, the statistical complexities
of grain-size distributions of alluvial sediment, and the
behavior of fluid flow and sediment transport in a large
meandering stream, suggest that a more sophisticated
approach to geomorphological studies of alluvial sedimentation is advisable. Conventional one-dimensional
steady-state hydraulic approaches have, in many cases, been
extended to their inherent limitations. Mean properties
of sediment and channel geometry likewise cannot explain
the variations in sediment-transport rates, flow velocity,

bedforms, and bed-material size, over short reaches in some streams. There is a keen need for a more complete understanding of the detailed interactions of fluid and sediment in these two-phase flows.

This philosophy of fluvial geomorphology means that difficult topics such as flow visualization, visualization of particle motions, spatial measurements of multidimensional velocity components, computational fluid dynamics, and mechanics of two-phase flow, must be pursued. One of the intellectually satisfying aspects of this pursuit is the avid exploration of these fields by scientists in other disciplines. The special manifestations of certain fluvial features provide an excellent opportunity for fluvial geomorphologists to contribute their own insights into the expansion of knowledge in these scientific topics.

Many topics in two-phase flow in alluvial channels which require sophisticated analysis were ignored in this paper but will be mentioned for the sake of completeness. Stochastic models of bedload transport (eg. Hung & Shen, 1971, 1976; Todorovic & Nordin 1975), stochastic models of the growth and evolution to steady state of ripples (Jain & Kennedy 1971, 1974), two- and three-dimensional stability analyses of flow-regime bedforms and of channel form (eg. Fredsøe 1974; Parker 1975, 1976), and analytical solutions of the dynamical equations of bend flow (Yen 1972; Engelund 1974) all involve methodologies that go far beyond conventional one-dimensional practices. At present, however, none of these four broad areas of engineering research have achieved results sufficiently explicit to be applicable to geomorphological problems.

ACKNOWLEDGEMENTS

The Department of Geology, University of Illinois at Urbana-Champaign, furnished field equipment for hydraulic surveys in the lower Wabash River. The Department of Geology, through the Shell Aids Program of the Shell Companies Foundation and the Geological Society of America Research Grant 1707-73, provided financial support for these efforts. Research during 1976 and 1977 was supported in part by NSF Grant EAR 76-10748. J. D. Milliman arranged the opportunity for measurements in the Fraser River. D. A. Vanko translated Grinvald (1965). H. Noll graciously allowed me to use the DuPont 310 Curve Resolver in his laboratory.

REFERENCES CITED

Ackers, P. & White, W.R., 1973, Sediment transport: new approach and analysis. *J. Hydraulics Division ASCE*, 99, 2041-2060

Allen, J.R.L., 1963, Henry Clifton Sorby and the sedimentary structures of sands and sandstones in relation to flow conditions. *Geologie en Mijnbouw*, 42, 223-228

Allen, J.R.L., 1964, Primary current lineation in the Lower Old Red Sandstone (Devonian), Anglo-Welsh Basin. *Sedimentology*, 3, 89-108

1. Sediment transport in alluvial streams

Allen, J.R.L., 1966, On bedforms and paleocurrents. *Sedimentology*, 6, *153-190*

Allen, J.R.L., 1968, *Current ripples*. North-Holland, Amsterdam, *433*

Allen, J.R.L., 1969, On the geometry of current ripples in relation to stability of fluid flow. *Geografiska Annaler*, 51A, *61-96*

Allen, J.R.L., 1970a, *Physical processes of sedimentation*. American Elsevier, New York, *248*

Allen, J.R.L., 1970b, A quantitative model of grain size and sedimentary structures in lateral deposits. *Geological Journal*, 7, *129-146*

Allen, J.R.L., 1970c, Studies in fluviatile sedimentation: a comparison of fining-upwards cyclothems, with special reference to coarse-member composition and interpretation. *J. Sedimentary Petrology*, 40, *298-323*

Allen, J.R.L., 1973, Phase differences between bed configuration and flow in natural environments, and their geological relevance. *Sedimentology*, 20, *323-329*

Amaral, E.J. & Pryor, W.A., 1977, Depositional environment of the St. Peter Sandstone deduced by textural analysis. *J. Sedimentary Petrology*, 47, *32-52*

Bagnold, R.A., 1941, *The physics of blown sand and desert dunes*. Methuen & Co. Ltd., London, *265*

Bagnold, R.A., 1966, An approach to the sediment transport problem from general physics. *US Geol. Survey Prof. Paper* 422-I, *37*

Blatt, H., Middleton, G. & Murray, R., 1972, *Origin of sedimentary rocks*, Prentice-Hall, Englewood Cliffs, New Jersey, *634*

Bogárdi, J.L., 1972, Fluvial sediment transport. *Advances in Hydroscience*, 8, *184-259*

Bridge, J.S., 1975, Computer simulation of sedimentation in meandering streams. *Sedimentology*, 22, *3-43*

Bridge, J.S. & Jarvis, J., 1976, Flow and sedimentary processes in the meandering River South Esk, Glen Cova, Scotland. *Earth Surface Processes*, 1, *303-336*

Briggs, L.I & Middleton, G.V., 1965, Hydromechanical principles of sediment structure formation. in: Middleton, G.V. (ed) *Primary sedimentary structures and their hydrodynamic interpretation*. Soc. Econ. Paleontologists and Mineralogists Special Publication, 12, *5-16*

Brush, L.M. Jr, Ho, H-W, & Yen, B.C., 1964, Accelerated motion of a sphere in a viscous fluid. *J. Hydraulics Division ASCE*, 90, *149-160*

1. Sediment transport in alluvial streams

Church, M. & Gilbert, R., 1975, Proglacial fluvial and lacustrine environments. In: Jopling, A.V. & McDonald, B.C. (eds), *Glaciofluvial and glacio-lacustrine sedimentation*. Soc. Econ. Paleontologists and Mineralogists Special Publication, 23, 22-100

Coleman, J.M., 1969, Brahmaputra River: channel processes and sedimentation. *Sedimentary Geology*, 3, 129-239

Costello, W.R., 1974, *Development of bed configurations in coarse sands*. Rept. 74-1, Earth & Planetary Science Dept, MIT, Cambridge, Mass, 120pp

Einstein, H.A., 1950, The bedload function for sediment transportation in open channel flows. *US Dept. Agriculture Tech. Bull*, 1026, 70pp

Einstein, H.A. & Li, H., 1958, Secondary currents in straight channels. *Trans. American Geophysical Union*, 39, 1085-1088

Einstein, H.A. & Shen, H.W., 1964, A study on meandering in straight alluvial channels. *J. Geophysical Research*, 69, 5239-5247

Engelund, F., 1974, Flow and bed topography in channel bends. *J. Hydraulics Division ASCE*, 100, 1631-1648

Francis, J.R.D., 1973, Experiments on the motion of solitary grains along the bed of a water-stream. *Proc. Royal Society London, Series A*, 332, 443-471

Fredsøe, J., 1974, On the development of dunes in erodible channels. *J. Fluid Mechanics*, 64, 1-16

Freeman, W.E. & Visher, G.V., 1975, Stratigraphic analysis of the Navajo Sandstone. *J. Sedimentary Petrology*, 45, 651-668

Galloway, W.E., 1976, Sediments and stratigraphic framework of the Copper River fan-delta, Alaska. *J. Sedimentary Petrology*, 46, 726-737

Gessler, J, 1971, Beginning and ceasing of sediment motion. In: Shen, H.W. (ed), *River mechanics, 1*, H. W. Shen, Fort Collins, Colorado, 7-1 to 7-22

Graf, W.H., 1971, *Hydraulics of sediment transport*. McGraw-Hill, New York, 513pp

Grass, A.J., 1970, Initial instability of fine sand bed. *J. Hydraulics Division ASCE*, 96, 619-632

Grass, A.J., 1971, Structural features of turbulent flow over smooth and rough boundaries. *J. Fluid Mechanics*, 50, 233-255

Grinvald, D.I., 1965, Some patterns of large-scale turbulence in flows. *Bull. USSR Academy Science, Geographic Series*, 3, 89-94 (in Russian).

Harms, J.C. & Fahnestock, R.K., 1965, Stratification, bed forms and flow phenomena (with an example from the Rio Grande). In: Middleton, G.V.(ed), *Primary sedimentary structures and their hydrodynamic interpretation*. Soc. Econ. Paleontologists and Mineralogists Special Publication, 12, 84-115

1. Sediment transport in alluvial streams

Heathershaw, A.D., 1974, "Bursting" phenomena in the sea. *Nature*, 248, 394-395

Heathershaw, A.D., 1976, Measurements of turbulence in the Irish Sea benthic boundary layer. In: McCave, I.N. (ed), *The benthic boundary layer*, Plenum Press, New York, 11-31

Hung, C.S. & Shen, H.W., 1971, Research in stochastic models for bed-load transport. In: Shen, H.W. (ed), *River Mechanics, 2*, H. W. Shen, Ft. Collins, Colorado, B-1 to B-47

Hung, C.S. & Shen, H.W., 1976, Stochastic models of sediment motion on flat beds. *J. Hydraulics Division ASCE*, 102, 1745-1759

Inman, D.L., 1949, Sorting of sediments in the light of fluid mechanics. *J. Sedimentary Petrology*, 19, 51-70

Jackson, R.G.II, 1975a, Hierarchial attributes and a unifying model of bed forms composed of cohesionless material and produced by shearing flow. *Bull. Geological Society America*, 86, 1523-1533

Jackson, R.G.II, 1975b, Velocity - bed-form - texture patterns of meander beds in the lower Wabash River of Illinois and Indiana. *Bull. Geological Society America*, 86, 1511-1522

Jackson, R.G.II, 1975c, *A depositional model of point bars in the Lower Wabash River*. PhD Dissertation, Univ. of Illinois, Urbana, 269pp

Jackson, R.G.II, 1976a, Largescale ripples of the Lower Wabash River. *Sedimentology*, 23, 593-623

Jackson, R.G.II, 1976b, Sedimentological and fluid-dynamic implications of the turbulent bursting phenomenon in geophysical flows. *J. Fluid Mechanics*, 77, 531-560

Jackson, R.G.II, 1976c, Unsteady-flow distributions of hydraulic and sedimentologic parameters across meander bends of the lower Wabash River, Illinois-Indiana, USA. *Proc. International Symposium Unsteady Flow in Open Channels, 12-15 April 1976, British Hydromechanics Research Association, paper G4*, 14pp

Jain, S.C. & Kennedy, J.F., 1971, The growth of sand waves. *Proc. International Symposium on Stochastic Hydraulics*, Pittsburgh University Press, 449-471

Jain, S.C. & Kennedy, J.F., 1974, The spectral evolution of sedimentary bed forms. *J. Fluid Mechanics*, 63, 301-314

Kalinske, A.A., 1943, The role of turbulence in river hydraulics. *Univ. Iowa Studies in Engineering Bull.*, 27, 266-279

Kline, S.J. Reynolds, W.C., Schraub, F.A. & Rundstadler, P.W., 1967, The structure of turbulent boundary layers. *J. Fluid Mechanics*, 30, 741-773

1.   Sediment transport in alluvial streams

Knighton, A.D., 1975, Variations in at-a-station hydraulic
    geometry. *American J. Science,* 275, 186-218

Kovasznay, L.S.G., 1967, Structure of the turbulent boundary
    layer. *Physics Fluids Suppl.,* 10, S25-S30

Laufer, J., 1975, New trends in experimental turbulence
    research. *Annual .Rev. Fluid Mechanics,* 7, 307-326

Leliavsky, S., 1955, *An introduction to fluvial hydraulics.*
    Constable, London, 257pp

Leopold, L.B. & Maddock, T., 1953, The hydraulic geometry
    of stream channels and some physiographic implications.
    US Geological Survey Prof. Paper 252, 57pp

Leopold, L.B., Wolman, M.G. & Miller, J.P., 1964, *Fluvial
    processes in geomorphology.* W. H. Freeman, San
    Francisco, 522pp

McQuivey, R.S., 1973, Summary of turbulence data from
    rivers, conveyance channels, and laboratory flumes.
    US Geological Survey Prof. Paper 802-B, 66pp

Matthes, G.H., 1947, Macroturbulence in natural stream
    flow. *Trans. American Geophysical Union,* 28, 255-262

Middleton, G.V.(ed), 1965, *Primary sedimentary structures
    and their hydrodynamic interpretation.* Soc. Econ.
    Paleontologists and Mineralogists Special Publication
    12, SEPM, Tulsa, Oklahoma, 265pp

Middleton, G.V., 1976, Hydraulic interpretation of sand
    size distributions. *J. Geology,* 84, 405-426

Moss, A.J., 1972, Bed-load sediments. *Sedimentology,* 18,
    159-220

Müller, A., 1976, Effect of secondary flow on turbulence
    in an open channel flow. *Proc. Second International
    IAHR Symposium on Stochastic Hydraulics,* 2-4 August
    1976, 21pp

Murray, S.P., 1970, Settling velocities and vertical
    diffusion of particles in turbulent water. *J.
    Geophysical Research,* 75, 1647-1654

Nychas, S.G., Hershey, H.C. & Brodkey, R.S., 1973, A
    visual study of turbulent shear flow. *J. Fluid
    Mechanics,* 61, 513-540

Offen, G.R. & Kline, S.J., 1974, Combined dye-streak and
    hydrogen-bubble visual observations of a turbulent
    boundary layer. *J. Fluid Mechanics,* 62, 223-239

Offen, G.R. & Kline, S.J., 1975, A proposed model of the
    bursting process in turbulent boundary layers.
    J. Fluid Mechanics, 70, 209-228

Onishi, Y., Kennedy, J.F. & Jain, S.C., 1972, *Effects of
    river curvature on a resistance to flow and sediment
    discharges of alluvial streams.* Iowa Institute
    Hydraulic Research, Univ. Iowa, Rept. Proj. A-029-IA,
    150pp

Oser, R.K., 1972, Sedimentary components of Northwest
    Pacific pelagic sediments. *J. Sedimentary Petrology,*
    42, 461-467

1.   Sediment transport in alluvial streams

Owen, P.R., 1964, Saltation of uniform grains in air. *J. Fluid Mechanics,* 20, 225–242

Parker, G., 1975, Sediment inertia as cause of river anti-dunes. *J. Hydraulics Division ASCE,* 101, 211–221

Parker, G., 1976, On the cause and characteristic scales of meandering and braiding in rivers. *J. Fluid Mechanics,* 76, 457–480

Rao, K.N., Narasimha, R. & Badri Narayanan, M.A., 1971, The "bursting" phenomenon in a turbulent boundary layer. *J. Fluid Mechanics,* 48, 339–352

Raudkivi, A.J., 1976, *Loose boundary hydraulics,* 2nd edition. Pergamon, Oxford, 397pp

Reed, W.E., Le Fever, R. & Moir, G.J., 1975, Depositional environment interpretation from settling-velocity (psi) distributions. *Bull. Geological Society America,* 86,· 1321–1328

Roshko, A., 1976, Structure of turbulent shear flows: a new look. *AIAA Journal,* 14, 1349–1357

Rubey, W.W., 1933, Settling velocities of gravel, sand and silt particles. *American J. Science, Series 5,* 25, 325–338

Saunderson, H.C., 1977, Grain size characteristics of sands from the Brampton esker. *Z. fur Geomorphologie,* 21, 44–56

Simons, D.B. & Richardson, E.V., 1961, Forms of bed rough-ness in alluvial channels. *J. Hydraulics Division ASCE,* 87 (HY3), 87–105

Simons, D.B., Richardson, E.V. & Nordin, C.F. Jr., 1965, Sedimentary structures generated by flow in alluvial channels. In: Middleton, G.V. (ed), *Primary sedi-mentary structures and their hydrodynamic interpretation.* Soc. Econ. Paleontologists and Mineralogists Special Publication 12, 34–52

Smith, J.D., 1969, Part 2: studies of nonuniform boundary-layer flows. In: *Investigations of turbulent boundary layers and sediment-transport phenomena as related to shallow marine environments.* Report A69-7, Oceanography Dept., Univ. Washington, Seattle

Smith, J.D. & McLean, S.R., 1977a, Spatially averaged flow over a wavy surface. *J. Geophysical Research,* 82, 1735–1746

Smith, J.D. & McLean, S.R., 1977b, Boundary layer adjustments to bottom topography and suspended sediment. *J. Geophysical Research,* 82 (in press)

Soo, S.L., 1967, *Fluid dynamics of multiphase systems.* Blaisdell, Waltham, Massachusetts, 524pp

Sternberg, R.W., 1972, Predicting initial motion and bed-load transport of sediment particles in the shallow marine environment. In: Swift, D.J.P., Duane, D.B. & Pilkey, O.H. (eds), *Shelf sediment transport: process and pattern.* Dowden, Hutchinson & Ross, Stroudsburg, Pennsylvania, 61–82

1.   Sediment transport in alluvial streams

Sundborg, A., 1956, The River Klarälven, a study in fluvial processes. *Geografiska Annaler,* 38, 125-316

Teleki, P.G., 1972, Wave boundary layers and their relation to sediment transport. In: Swift, D.J.P., Duane, D.B. & Pilkey, O.H. (eds), *Shelf sediment transport: process and pattern.* Dowden, Hutchinson & Ross, Stroudsburg, Pennsylvania, 21-59

Todorovic, P. & Nordin, C.F.Jr., 1975, Evaluation of stochastic models describing movement of sediment particles on river beds. *J. Research US Geological Survey,* 3, 513-517

USACE, 1950, Turbulence in the Mississippi River, US Army Corps of Engineers, *Waterways Experiment Station, Potamology Investigation, Rept* 10-2, 32pp

Van Andel, Tj. H. 1973, Texture and dispersal of sediments in the Panama Basin. *J. Geology,* 81, 434-457

Vanoni, V.A., 1975, River dynamics. *Advances in Applied Mechanics,* 15, 1-87

Visher, G.S., 1969, Grain size distributions and depositional processes. *J. Sedimentary Petrology,* 1074-1106

Visher, G.S. & Howard, J.D., 1974, Dynamic relationship between hydraulics and sedimentation in the Altamaha estuary. *J. Sedimentary Petrology,* 44, 502-521

Waldrop, W.R. & Tatom, F.B., 1976, Analysis of the thermal effluent from the Gallatin steam plant during low river flows. *Tennessee Valley Authority Report* 33-30, Norris, Tennessee

White, W.R., Milli, H. & Crabbe, A.D., 1975, Sediment transport theories: a review. *Proc. Institution Civil Engineers,* Pt. 2, 59, 265-292

Yalin, M.S., 1972, *Mechanics of sediment transport.* Pergamon Press, Oxford, 290pp

Yalin, M.S., 1975, On the development of sand waves in time. *Proc. XVI Congress International Association Hydraulic Research,* 2, 212-219

Yalin, M.S., 1976, On the origin of submarine dunes. *Proc. XV Coastal Engineering Conference,* 9pp

Yang, C.T., 1976, Discussion of "Sediment transport theories: a review". *Proc. Institution Civil Engineers* Pt. 2, 61, 803-810

Yang, C.T. & Stall, J.B., 1976, Applicability of unit stream power equation. *J. Hydraulics Division ASCE,* 102, 559-568

Yen, B.C., 1972, Spiral motion of developed flow in wide curved open channels. In: Shen, H.W.(ed), *Sedimentation,* H. W. Shen, Ft. Collins, Colorado, 22-1 to 22-33

Yen, B.C., 1973, Open-channel flow equations revisited. *J. Engineering Mechanics Division ASCE,* 979-1009

1. Sediment transport in alluvial streams

Yen, C-L., 1970, Bed topography effect on flow in a
    meander. *J. Hydraulics Division ASCE,* 96, 57–73

Zelazny, S.W. & Baker, A.J., 1975, Predictions in
    environmental hydrodynamics using the finite element
    method, II, applications. *AIAA Journal,* 13, 43–46

2 ASPECTS OF FLOW RESISTANCE IN STEEP CHANNELS HAVING COARSE PARTICULATE BEDS

*T. J. Day

ABSTRACT

A review of experimental flow resistance data for sub-critical flows in steep, coarse bed-material channels indicates a radical departure from conventional formulae for low values of relative roughness. A unique resistance function does not exist for values of relative roughness less than about 3, and limited data are presented which indicate that roughness characteristics dominate flow properties. Coherent relationships between flow resistance, bed configuration and surface deformation are not as yet defined.

INTRODUCTION

Flow properties of streams in mountain areas are receiving increased attention as environmental concern for the maintenance of natural waterways expands, and engineering concern for the design and preservation of transportation corridors continues. Quite obviously a wide variety of mountain stream types exist in response to the imposed runoff regime, nature of sediment supply, and in some cases the recent evolutional history of the drainage basin (eg. in some recent deglaciated basins the channel slope is imposed). The following discussion, however, is limited to rough, high-gradient channels, with discrete bed elements which extend through a major portion of the stream depth and whose slope is so great that significant disturbances will form over the bed elements under some stages of discharge.

Even within the limits of this restriction a significant range of flow properties exist. For example, in a series of experimental studies at Utah State University (Mohanty 1959; Mirajgoaker 1961; Al-Khafaji 1961; Kharrufa 1962; Judd 1963; Hariri 1964) flow properties in these channels were shown to vary with bed configuration, slope and discharge. As no single equation or definition fitted all the range of variation, several types or 'regimes' were identified: 1) the tranquil regime which includes all subcritical flow having low Froude numbers

* Terrain Sciences Division, Geological Survey of Canada, 601 Booth Street, Ottawa, Ontario, Canada

2. Flow resistance in steep channels

and slopes; 2) the tumbling regime where hydraulic jumps
dissipate stream energy rather than through boundary
friction; 3) the rapid regime where supercritical flow
predominates. Several authors (eg. Mohanty 1959;
Gordienko 1967) also suggest a fourth transitional regime
between tumbling and rapid. For these experimental studies
each regime can be associated with ranges or limits of
Froude number and channel slope (cf. Mohanty 1959, pp 81-
82); however, this criteria cannot be directly applied to
natural streams due to complexities in channel and flow
structure.

The type of flows considered herein are confined to
sub-critical conditions with Froude numbers (up to 0.9)
which exceed the criteria for the tranquil regime, whereas
the data are all within the experimentally derived slope
criteria of 3.2%. The purpose of this discussion is to
review various experimental studies with a view to co-
ordinating their results into a statement of where further
work is required.

FLOW RESISTANCE CONSIDERATIONS

In the most general case, flow resistance is a function of:
Reynolds number (ratio of inertial to viscous forces);
relative roughness (refers to the ratio of flow depth to
roughness height; however, the spacing, shape and size
distributional properties of the bed materials must also
be considered); cross-sectional shape, channel non-
uniformity (variations in planform geometry and large
scale bed deformations); Froude number (ratio of inertial
to gravitational forces), and the degree of unsteadiness.
The first three variables describe the effect of surface
resistance, the next two describe the additional effect of
form and wave resistance and the last is associated with
local acceleration.

Resistance equations are derived by balancing the
resistance shear force at the channel boundary with the
propelling force. Formulation of this relationship
requires, however, certain assumptions concerning the
structure of the velocity distribution, flow type and
channel cross-section. For example, the velocity distribution in a two-dimensional open channel flow over a rough
boundary (roughness height is much larger than the viscous
sublayer thickness) usually follows a logarithmic law, and
when integrated across the flow depth leads to the equation
for the Darcy friction factor, f:

$$\frac{1}{\sqrt{f}} = c_o + 2.03 \log \frac{d}{k} \tag{1}$$

where c_o is a constant, d is the distance from the
boundary, k is a roughness height, and d/k is a relative
roughness factor. The value of c_o in Equation (1) can be
determined from the size of distributional characteristics
of the bed material (Burkham and Dawdy, 1976). Equation
(1) also can be expressed in terms of the ratio of the
mean cross-section velocity v, to the bed shear stress, u_*

2. Flow resistance in steep channels

as follows:

$$\frac{v}{v} = c_1 + 5.75 \log \frac{d}{k} \qquad (2)$$

where c_1 is another constant. An empirical formula used extensively in open channel flow problems is the Manning Equation:

$$v = R^{2/3} S^{1/2} / n \qquad (3)$$

where R is the hydraulic radius, S is slope and n a total roughness coefficient which may be determined from grain size data, i.e. the Sticker equation. In terms of the ratio v/v_*, Manning's Equation can be manipulated into a power law form

$$\frac{v}{v_*} = c_2 \left(\frac{d}{k}\right)^{c_3} \qquad (4)$$

where c_2 and c_3 are constants.

Application of these resistance equations to natural channels has met with mixed success due to the existence of nonuniform boundaries and patterns. In natural channels, stream energy is dissipated by particle resistance but also by bedforms, channel curvature and excessive deformation of streamlines (spill resistance). However, for natural gravel channels where d/k is large, Equation (2) and (4) offer an adequate representation (Kellerhals 1967; McDonald and Lewis 1973) of the friction relationship.

The application of these equations to rough, high-gradient channels is incorrect for several reasons. As roughness increases the assumption of logarithmic velocity distribution becomes invalid. O'Loughlin and Annambhotla (1968) found that a departure from the logarithmic law occurs near the roughness elements, presumably due to the presence of wakes behind the isolate cubes used in the experiments. This break in the velocity profiles occurred at a depth approximately twice the height of the roughness elements. Velocity distributions over naturally sorted beds composed of heterogeneous particles with the larger roughness heights protruding well into, and at some stages through the flow, are exceedingly complex and such flows cannot be adequately described by Equation (1).

The presence of surface deformations associated with steep slopes also precludes application of uniform flow formulaes. Ashida *et al* (1973) have shown that for $0.4 \leqslant d/k \leqslant 2.0$ (k based on a uniform sphere diameter) flow resistance decreases as channel slope increased for any constant value of d/k, and attributed this to the effect of surface waves. For flows over natural beds where irregular surface waves occur, neither the distribution law for the velocity or the characteristics of the turbulent flow are presently evident.

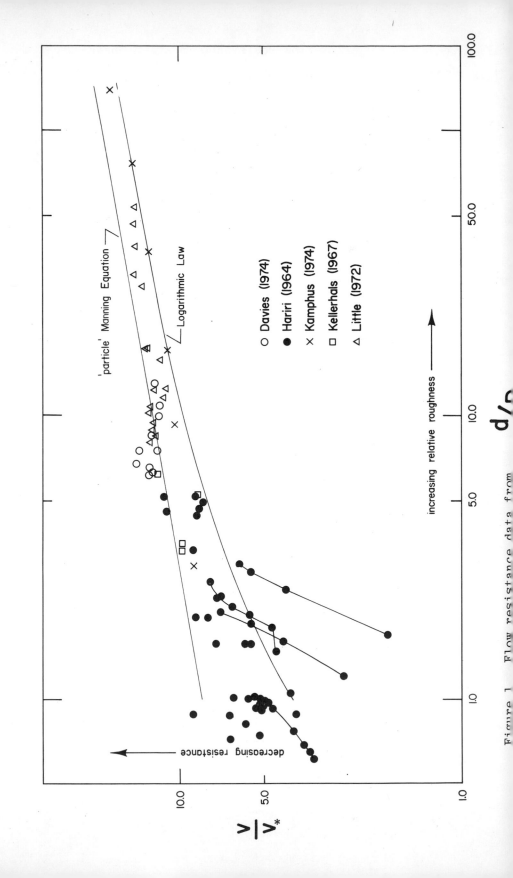

Figure 1 Flow resistance data from

2. Flow resistance in steep channels

DATA FROM EXPERIMENTAL STUDIES

A compilation of data from five experimental flume studies of flow properties over particulate beds is illustrated in Figure 1. The roughness height parameter, k, is estimated by D_{90}, which is the size below which 90% of the bed material occurs. This procedure leads to a bias towards the large grains on the bed which, according to various observations (eg. White 1940) account for a significant portion of the resistance. Due to nonequivalent sampling procedures (Little 1972 and Davies 1974, D_{90} based on surface layer only; Hariri 1964, D_{90} base on first two layers of material; Kellerhals 1967, D_{90} base on total bed material) the data shown in Figure 1 are not completely compatible; however, the differences are assumed to be small in comparison to variations occurring from other sources discussed below.

The logarithmic law plotted on Figure 1 has a value of 6.2 for c_0. This value was arrived at by using the method outlined by Burkham and Dawdy (1976). The Manning Equation is also shown with $c_2 = 8.4$ and $c_3 = 0.17$ (Equation 4).

Inspection of Figure 1 leads to the observation that both the Mannings Equation and the logarithmic law offer an acceptable representation for experimental flows where d/k ($k = D_{90}$) is larger than about 3. There also appear to be no difference in the resistance properties of naturally developed particular beds (Davies, Kellerhals & Little) and artificial beds composed of natural materials. Kamphus (1974) formed beds by glueing grains (0.5 mm $<$ D $<$ 38 mm) to a vinyl paper.

Below $d/D_{90} = 3$ departures from Equation (1) and (4) become more noticeable. In point of fact Hariri's data for experimental tests of several flows measured over the same slope and natural bed roughness characteristics (indicated by the joined points in Figure 1) suggest a significantly different form for the resistance equation with high initial resistances, and a much more rapid decrease as flow scale increases. Closer inspection of these specified data suggest that the friction relationship changes with both the bed material characteristics and channel slope. Data for flows over a coarse bed material (220 mm) plot to the left of flows over a small material ($D_{90} = 90$ mm) indicating a higher resistance for the smaller sized bed material for any value of d/D_{90} between 0.6 and 3.0. This rather contradicting observation is presumably the result of different packing characteristics with the bed formed of 220 mm size material presenting a more uniform boundary. Hariri's data also indicate that for constant bed roughness characteristics an increase in channel slope shifts the relationship to the left for $D_{90} = 90$ mm material (indicating less resistance), supporting Ashida's *et al* (1973) observations. However, for the coarser sized material a change of bed slope from 0.02 to 0.04 does not result in any shifting of the data.

The results of a series of experiments undertaken by the author are shown in Figure 2. The bed was composed of

2. Flow resistance in steep channels

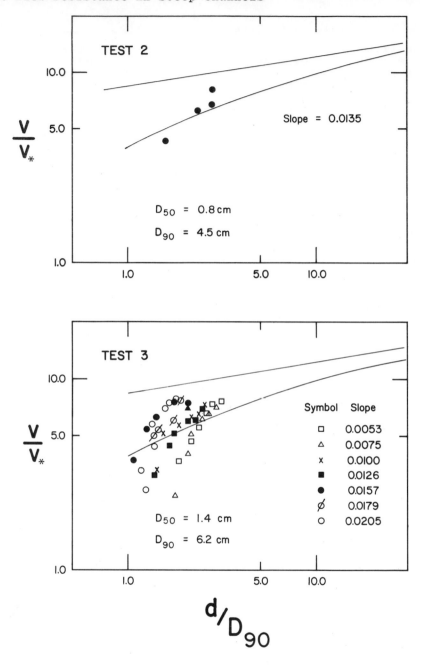

Figure 2   Flow resistance data from
            four experiments under-
            taken by the author

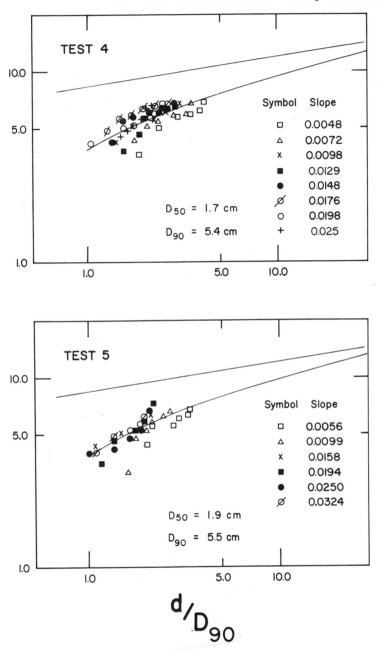

Figure 2   Continued

2.   Flow resistance in steep channels

natural fluvial material ranging from 2 to 90 mm (the size distribution was truncated at 2 mm to prevent re-circulation of sands).   Each successive test bed was sub-mitted to increased bed shear stress to remove the finer size fractions, and the stabilized bed thereby formed, studied for its resistance properties in the same way as done by Hariri.   All flows were in the subcritical range.

Results from Test 2 and 3 indicate a friction relation-ship similar to that shown for Hariri's data, with a high rate of change in resistance properties as flow scale increases.   Test 3 data also show the trend for decreasing resistance as channel slope increases (bed roughness characteristics are the same for all measurements).   As the rate of change of resistance does not vary significantly among the seven data sets for Test 3 it is possible to speculate that the roughness characteristics of the bed are the determining factor, whereas channel slope is in some way related to the value of v/v_* for any specific d/D .

In Test 4 further degradation of the bed occurred until another stable armour coat developed.   As seen in Figure 2 the new bed had significantly different friction responses. For value of d/D_{90} of approximately 3 and larger, the data appear to follow the logarithmic law, but for $d/D < 3$ the rate of change in resistance increases, and the responses begin to diverge as in Test 3.   Further experimentation resulted in Test 5 which shows little difference from the resistance relationships of Test 4.   A photograph of the Test 5 bed is shown in Figure 3.

Examples of the influence of roughness characteristics are shown in Figure 4 where the resistance data for similar slopes but different bed roughness characteristics are shown.   Increased armouring produced a channel bed of more resistance.   This observation is an important one for consideration in modelling natural flows having lower values of d/k.

Data presented in Figure 1 and 2 show two important considerations: 1) that conventional resistance equations are invalid for $d/k < 3$, presumably due to nonuniform velo-city distributions and surface distortions;   2) that for values of $d/k < 3$ flow properties are dominated by rough-ness characteristics, although exactly which characteris-tics and in what manner remains to be proven.

Present data offer some insight into this problem of dominant roughness characteristics.   Peterson's and Mohanty's (1960) flume studies of flow in steep, rough channels (modelled by arrays of bar and cube elements) indicated that the most important parameter representing roughness geometry may be the roughness concentration, defined as the ratio of the projected effective area of the roughness elements perpendicular to the direction of flow to the total area of the channel bed.   More recently Williman (1975) speculated that an armoured stream bed may possess a dominant roughness element at long wave lengths (approximately 20 times the mean stone width for his data), and also found that the concentration of particles had an

Figure 3   The armoured bed formed in Test 5

2. Flow resistance in steep channels

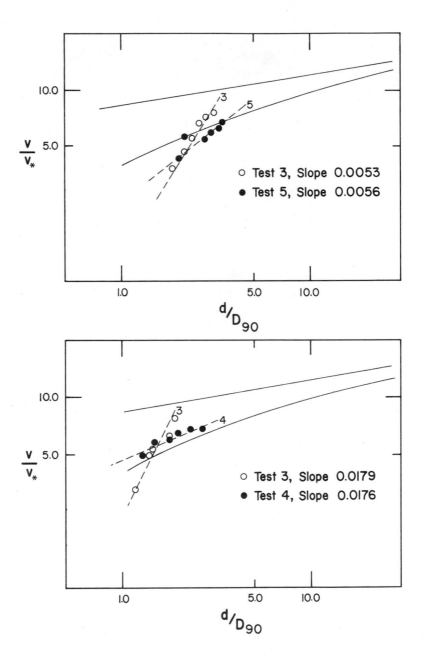

Figure 4   Diagram showing the effect of increased armouring on the resistance properties of a gravel bed.

average of 0.19. Rouse (1965) determined that the optimum concentration for maximum roughness was with areal concentrations between 0.15 and 0.25, so Williman's results suggest that an armoured bed surface presents the maximum hydraulic roughness for their size.

Data presented in Figure 4 are also from armoured beds (formed through winnowing away of fines and the stabilization of residual particles, a process typical to mountain streams). Although areal concentration data are not available for these experiments some indicators on the change in roughness height characteristics are shown in Figure 5. The height frequencies shown here for the original bed and Tests 2, 3 and 4 are based on heights above mean bed height over a one meter distance. Changes in frequency characteristics are more clearly shown in the composite plot of cumulative curves. Comparing Tests 3 and 4 it is evident that even though D_{90} actually decreased from 6.2 to 5.4 cm (cf. Figure 2) the distribution of roughness heights are quite different, with large roughness heights for Test 4 and the friction relations similarly different. Unfortunately without knowing the relationship between height frequencies and areal concentrations it is not possible to say what the data in Figure 2 and 5 mean in terms of Williman's observation.

Data presented herein are certainly insufficient to make any conclusion with the exception that for d/k >3 the logarithmic flow law (k = D_{90}) offers an adequate representation of the resistance functions, and for d/k <3 such a unique relationship does not exist. Further information as to the characteristics and the effects of natural bed particle parameters (height, spacing, concentration), and the effects of surface deformations, are required.

REFERENCES CITED

Al-Khafaji, A.N., 1961, *The dynamics of two-dimensional flow in steep, rough, open channels.* PhD dissertation, Utah State Univ., Logan, Utah

Ashida, K., Daido, A., Takahashi, T. & Mizuyama, T., 1973, Study of the resistance law and the initiation of motion of bed particles in a steep slope channel. *Annual Report 16B, Disaster Prevention Laboratory, Kyto Municipal Laboratory*

Burkham, D.E. & Dawdy, D.R., 1976, Resistance equation for alluvial-channel flow. *J. Hydraulics Division ASCE,* Hy10, October, 1479-1489

Davies, B.E., 1974, *The armouring of alluvial channel beds.* ME thesis, Univ. of Canterbury, Christchurch, New Zealand.

Gordienko, P.I., 1967, The influence of channel roughness and flow states of hydraulic resistances in turbulent flow. *J. Hydraulic Research,* 5(4), 247-261

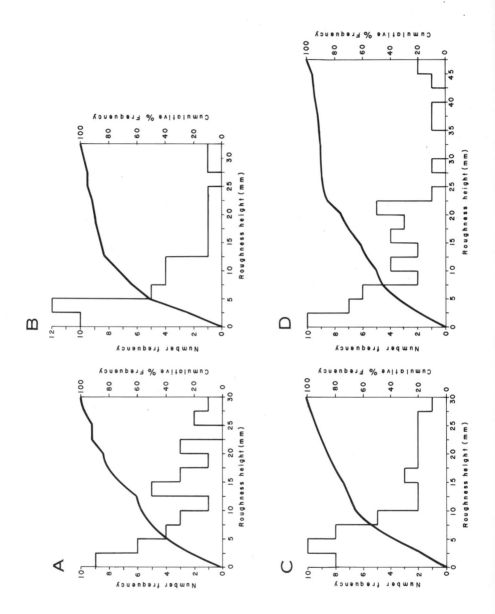

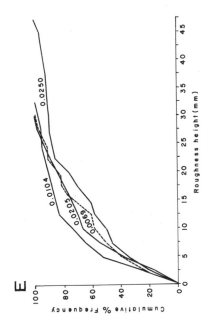

Figure 5
Roughness height frequencies for the
initial bed (A), Test 1(B), Test 2
(C), Test 3(D). A composite plot of
cumulative frequencies is shown in E,
where each curve is identified by the
stabilizing slope, starting with the
initial bed at 0.0069 to Test 4 at
0.0250

2. Flow resistance in steep channels

Hariri, D., 1964, *Relation between the bed pavement and the hydraulic characteristics of high-gradient channels in noncohesive sediments.* PhD Dissertation, Utah State Univ., Logan, Utah

Judd, H.E., 1963, *A study of bed characteristics in relation to flow in rough high-gradient, natural channels.* PhD Dissertation, Utah State Univ., Logan, Utah

Kamphuis,J.W., 1974, Determination of sand roughness for fixed beds. *J. Hydraulic Research,* 12(2), 193-203

Kellerhals, R., 1967, Stable channels with gravel-paved beds. *J. Waterways and Harbours, ASCE,* 93(WW1), 63-84

Kharrufa, N.F., 1962, *Flume studies of flow in steep, open channels with large graded, natural roughness elements.* PhD Dissertation, Utah State Univ., Logan, Utah

Little, W., 1972, *The role of sediment gradation on channel armouring.* PhD Dissertation, Georgia Institute of Technology, Atlanta, Georgia

McDonald, B.C. & Lewis, C.P., 1973, *Geomorphic and sedimentologic processes of rivers and coasts, Yukon Coastal Plain.* Canada, Environmental Social Comm. Northern Pipelines, Task Force on Northern Oil Development, Report 73-74

Mirajgoaker, A., 1961, *Effects of single large roughness elements in open channel flow.* PhD Dissertation, Utah State Univ., Logan, Utah

Mohanty, P.K., 1959, *The dynamics of turbulent flow in steep, rough open channels.* PhD Dissertation, Utah State Univ., Logan, Utah

O'Loughlin, E.M. & Annambhotla, V.S.S., 1968, Flow phenomena near rough boundaries. *J. Hydraulic Research,* 7(2), 232-250

Peterson, D.F. & Mohanty, P.K., 1960, Flume studies of flow in steep, rough channels. *J. Hydraulics Division ASCE,* HY9, 55-76

Rouse, H., 1965, Critical analysis of open channel resistance. *J. Hydraulics Division ASCE,* HY4, 1-25

White, C.M., 1940, Equilibrium of grains on the bed of a stream. *Proc. Royal Society of London, Series A,* 174, 322-334

Williman, E.B., 1975. *The development of armoured surfaces in streambed channels.* ME Thesis, Univ. of Canterbury, Christchurch, New Zealand

3 HYDRAULIC FACTORS CONTROLLING CHANNEL MIGRATION

*Edward J. Hickin

ABSTRACT

Surveys of the Beatton River indicate that many of its meander loops have been formed by discontinuous migration phases. Each phase is terminated when a segment of channel develops a critical curvature ratio ($r_m/w = 2.0$), a phenomenon previously explained by the writer in terms of the Bagnold separation-collapse theory. Observations on markedly curved bends of the Beatton River indicate, however, that convex-bank separation does not occur there; flow separation instead occurs at the concave banks. Although concave-bank separation is not commonly observed in flume studies, its absence may be a consequence of the common but unreal assumption of constant channel width; there is some evidence of concave-bank separation in other river studies. This phenomenon provides a hydraulic explanation of arrested channel migration which is an alternative to the Bagnold separation-collapse theory. Specific types of further research are recommended and a possible engineering application of induced concave-bank separation is briefly outlined.

INTRODUCTION

Previous studies by Hickin (1974) and Hickin and Nanson (1975) have shown that, probably for freely-meandering rivers in general, but certainly for the particular case of the Beatton River in NE British Columbia, river migration is a markedly discontinuous process. Each floodplain segment contained by a meander loop of the Beatton River consists of sediments deposited during a number of distinct migration phases. Each phase in turn consists of an initiation stage, a growth period, and an abrupt termination stage.

The scroll-bar array of the Beatton River floodplain faithfully records this pattern of discontinuous channel-shifting (see Hickin, 1974), an example of which is illustrated in Figure 1.

Here phase 1 is marked by relatively slow migration from a slightly curved initial channel. Scroll-bar spacing suggests that, during the growth period, the migration rate reaches a maximum, after which it abruptly declines to zero at the termination stage. The termination stage occurs

*Department of Geography, Simon Fraser University, Burnaby, British Columbia, Canada V5A 1S6

3. Channel migration

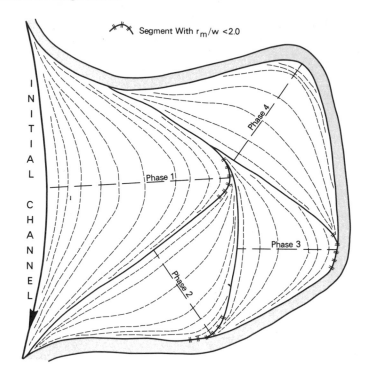

Segment With r_m/w <2.0

INITIAL CHANNEL

Phase 1

Phase 2

Phase 3

Phase 4

Figure 1   A complex meander loop produced by four distinct migration phases

when any segment of the channel develops a tightly-curved reach in which the radius of channel curvature, r_m is about twice the magnitude of the bank-full channel-width, w (see Figure 2).   Such channel segments are said to have achieved the critical curvature ratio (r_m/w = 2.0).

The arrested migration phase 1 is followed by the initiation stage of phase 2.   A new lobe of the floodplain begins to develop on the downstream limb of the phase 1 channel bend.   Phase 2 similarly displays a growth period of relatively rapid migration followed by an abrupt cessation of migration when the phase 2 channel achieves the critical curvature-ratio.   In this particular case the cycle is subsequently repeated through phases 3 and 4.

The migration-rate sequence from slow, to relatively rapid, to negligible values, during a single migration phase, was at first inferred from scroll-bar patterns on aerial photographs (Hickin 1974).   The existence of this sequence has been confirmed subsequently by detailed ground surveys of floodplain surface-form and of the age structure in the floodplain vegetation (Hickin & Nanson 1975).   The results of this field study, summarised in Figure 3, clearly accord with the pattern described above.

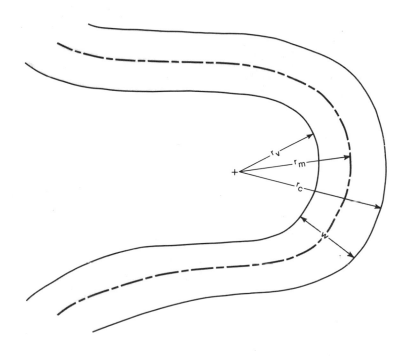

Figure 2   Elements of meander-planform geometry

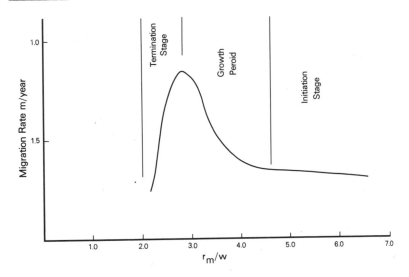

Figure 3   The relation of river migration-rate to
channel curvature during the initiation
stage, growth period and termination stage
of meander-loop development

THE BAGNOLD SEPARATION-COLLAPSE THEORY

The hydraulic mechanism underlying the correlation of
minimum migration-rate and critical curvature ratio is not
known.   It has been suggested, however, that the Bagnold
separation-collapse theory (Bagnold 1960) may provide a
hydraulic explanation of the character of the termination
stage (Hickin 1974).   Bagnold argued that the reduction in
flow resistance observed to occur in some channel bends of
decreasing curvature-ratio, could be explained by the
development of a strong separation zone at the convex bank.
The separation zone effectively narrows the channel and
increases the overall hydraulic efficiency of the bend.
Bagnold further argued that the rapid increases in flow
resistance observed in very tight bends (where $r_m/w < 2.0$)
resulted from the eventual collapse and disintegration of
the same separation zone.   This disintegration is accompanied by the generation of large-scale eddies at the
separation boundary and thus by markedly increased turbulence, by complete disruption of the flow pattern, and by
a consequent rapid increase in flow resistance.   Prior to
the collapse of the separation zone the concave bank supports considerable boundary shear from the high-velocity
filament of the flow;   after the collapse it is likely that
velocity and shear stress (and erosive capacity) are
markedly reduced at the bank.

The separation-collapse theory is a particularly
appealing explanation of the abrupt reduction in migration
rate in the termination phase because it is itself a dis-
continuity phenomenon representing a threshold condition in
the flow.

The character of separation zones at the convex banks
of channels has been described in many flume studies (for
examples see Mockmore 1943; Rozovskii 1957;   Bagnold 1960;
Ippen & Drinker 1962;   Hooke 1975).   Similar detailed descriptions of separation zones in natural channels, on the
other hand, are virtually unknown.

VELOCITY DISTRIBUTION IN
BEATTON RIVER BENDS

Velocity distributions in several cross-sections at very
tight bends of the Beatton River were recorded at near bankfull flow in the summer of 1975.   The aim of this study was
to document the character of separation-zone formation and
its subsequent collapse at the convex bank.   However, not
only was the collapse unrecorded, but there was also no
evidence of the formation of a strong separation zone at
the convex bank.   On the contrary, a definite shift in the
high-velocity filament of flow towards the convex bank
maintained that part of the channel as a very active flow
and sediment-transport zone (a phenomenon also noted elsewhere by others;   for example see Mathes 1951, and Hooke
1975).   At the concave bank, on the other hand, a separation
envelope was observed to develop and isolate that segment
of the channel from the main flow.

EXPERIMENTAL STUDIES AND
THE BOUNDARY ASSUMPTIONS

These rather surprising observations are, of course, contrary to those recorded for most experimental flume channels. A comparison of the boundary conditions adopted in flume studies to those existing in real channels, may provide an explanation of these apparently anomalous results.

Invariably experimental channels are constructed with a constant width through the channel bend. This seemingly reasonable approximation of reality is in fact quite misleading.

If we consider such a bend (Figure 2), it is obvious from the geometry that the radius of curvature at the concave bank (r_c) must always be greater than that at the convex bank(r_v). Furthermore, as the bend tightens and the curvature ratio (r_m/w) declines, the difference between the two bank radii increases in accordance with

$$r_c/r_v = \frac{r_m + \frac{w}{2}}{r_m - \frac{w}{2}}$$

a relationship which is graphed in Figure 4. In such a model a channel of critical curvature ratio has a concave

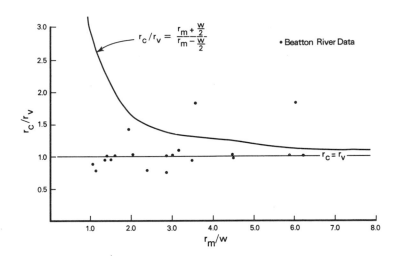

Figure 4   Hypothetical and actual (Beatton River) values of radii of curvature at the concave and convex banks (r_c/r_v) as curvature ratio (r_m/w) varies.

3.   Channel migration

bank radius which is 1.67 times as great as that at the convex bank. Clearly this type of geometry encourages separation at the convex bank of a tightening channel bend long before it is likely to occur at the concave bank.

In contrast, data from the Beatton River indicates no such increase in r_c/r_v as r_m/w declines (see Figure 4). In fact, several tight bends (r_m/w <3.0) known to be at the termination stage of migration, have an average value of r_c/r_v which is slightly less than unity (0.96).

These contrasting results are a direct consequence of violating the constant channel-width assumption. The channel width in tight bends of the Beatton River flares to above average values and gives rise to relatively small concave-bank radii.

FLOW SEPARATION AT THE CONCAVE BANK
AS A CONTROL ON CHANNEL MIGRATION

The evidence outlined above suggest an explanation of the termination stage of channel migration which is an alternative to Bagnold's separation-collapse theory.

As a developing channel bend approaches the critical curvature ratio, the migration rate reaches a maximum which is characterised, however, by a short-term disequilibrium between sedimentation at the convex bank and erosion at the concave bank. At this stage the concave bank, subjected to the intense inertial forces in the flow, erodes so rapidly that it widens the channel at the bend axis. This channel widening is not accompanied by general realignment of the channel upstream and downstream of the bend axis and very rapidly produces a small radius concave-bank segment and an associated separation zone. Once this separation envelope is established, erosion of the enclosed concave bank ceases. Although some sedimentation at the convex bank is likely to continue, the channel width will remain relatively large because of the reduced mean-velocity associated with the high degree of internal-distortion resistance in the bend.

GEOMORPHIC EVIDENCE OF CONCAVE-BANK
FLOW SEPARATION

Although the hydraulic evidence forming the basis of this proposed mechanism of migration control is obviously rather limited, there does appear to be additional supporting geomorphic evidence from elsewhere. Floodplain sediments of the lower Mississippi River are reported by Carey (1969) to be partly composed of "concave or eddy accretions". Deposition of the concave banks rarely occurs unless flow separation has occurred. Similar evidence was reported by Woodyer (1975) from a study of a "hairpin bend" on the Barwon River in New South Wales, Australia. He found that very active sedimentation of fine material occurred in a zone of reverse flow at the concave bank, forming "concave-bank benches".

CONCLUSIONS AND IMPLICATIONS
FOR FUTURE RESEARCH

I am not suggesting here that concave-bank flow-separation constitutes a universal explanation of hydraulically arrested channel-migration. However, there is convincing evidence that it is clearly the factor limiting the lateral movement of some channel bends. Furthermore, I believe that the lateral movement of any migrating channel, which is free to widen through bends, will probably be constrained by flow conditions at the concave rather than convex bank.

The character of this paper is in part speculative and there clearly is a need for two types of further research. The first should involve experimental work on flume channels with realistic boundary conditions. Bend configuration of real channel planform must provide the necessary basis of useful flume studies of flow through channel bends.

The second type of research, in my opinion the more important, should involve detailed measurements of the flow structure through channel bends of varying r_m/w. Only when such information is at hand can we hope for solutions to problems relating to channel migration.

If tightening of the concave-bank curvature does generally induce flow separation and thus protects that bank from further erosion, these ideas may be of particular interest to the engineer. Present engineering practise deals with laterally unstable channel bends by building the concave bank out into the channel and revetting the artificial surface. This method is a sometimes useful short-term measure but often fails over several years. It can easily be argued from the present work that a better solution would be provided by excavating the concave bank further and letting the resulting separation zone do the work of the revetment.

REFERENCES CITED

Bagnold, R.A., 1960, Some aspects of the shape of river meanders. *US Geological Survey Prof. Paper* 282-E, 135-144

Carey, W.C., 1969, Formation of flood plain lands. *American Society Civil Engineer Proc.,* HY3, 95, 981-994

Hickin, E.J., 1974, The development of meanders in natural river-channels. *American J. Science,* 274, 414-442

Hickin, E.J. & Nanson, G.C., 1975, The character of channel migration on the Beatton River, Northwest British Columbia, Canada. *Geological Society America Bull.,* 86, 487-494

Hooke, R.leB., 1975, Distribution of sediment transport and shear stress in a meander bend. *J. Geology,* 83, 543-565

3. Channel migration

Ippen, A.T. & Drinkwater, P.A., 1962, Boundary shear stresses in curved trapezoidal channels. *American Society Civil Engineers Proc.*, HY5, 88, 143-179

Mockmore, C.A., 1943, Flow around bends in stable channels. *American Society Civil Engineers Trans.*, 109, 335-360

Rozovskii, I.L., 1957, *Flow of water in bends of open channels*, *Academy of Sciences of the Ukranian SSR*. Institute Hydrology and Hydraulic Engineering (J. Program for Scientific Translations, 1961), 233pp

Woodyer, K.D., 1975, Concave-bank beaches on Barwon River, NSW. *Australian Geographer*, 13(1), 36-40

4 THE INTERPRETATION OF ANCIENT ALLUVIAL SUCCESSIONS IN THE LIGHT OF MODERN INVESTIGATIONS

*Brian R. Rust

ABSTRACT

A new classification of channel systems is proposed, based on the braiding parameter: the number of braids per mean meander wavelength. Single-channel and multi-channel systems have braiding parameters less than and more than one, respectively. Systems are further divided into low and high sinuosity at the boundary 1.5, giving four types, of which high sinuosity single-channel (meandering) and low sinuosity multi-channel (braided) are by far the most common.

The depositional model for meandering systems is well established: fining-upward cycles, with a predominantly sandy channel unit, and a fine overbank unit. It allows reasonable estimation of dimensions and hydrological parameters of paleochannels.

Braided deposits are divided into three categories: framework gravel-, sand-, and silt-dominated, the latter being rare. High slope proximal gravels (alluvial fans) are distinguished from low slope equivalents by greater lithological variation, more rapid downslope fining and (commonly) the presence of debris flow deposits. Both have in common an abundance of horizontally-stratified, imbricate, framework gravel, deposited chiefly on longitudinal bars. Contrary to the established view, these bars are regarded as equilibrium bedforms stable under flood conditions capable of moving all the bed material. Distal gravels are characterised by fining-upward sequences dominated by large-scale trough cross-beds of framework gravel, but also have fine units formed on inactive braided areas.

Proximal sandy braided deposits show great lateral and vertical variability, with abundant erosional surfaces and mudstone intraclasts, but rare primary mudstones. In contrast, distal sandy deposits have lateral continuity and repetitive sequences, and are transitional into deposits of meandering systems. Attempts have been made to estimate dimensions and hydrologic parameters of ancient braided systems, but at present they are hampered by insufficient knowledge of modern braided rivers.

*Department of Geology, University of Ottawa, Ottawa, Ontario, Canada K1N 6N5

4. Ancient alluvial successions

INTRODUCTION

The advantages of using modern alluvial systems for interpreting their ancient equivalents have been recognised
since the time of Hutton's famous "Theory of the Earth",
presented in 1785. The modern environment and its micro-
varieties can be accurately described and diagnosed, and
in most cases its processes and products can be quantified.
The advantages are not all on one side, however, for ancient
successions can provide extensive exposures normal to the
depositional surface, lacking in modern environments. In
addition, stratigraphic successions commonly preserve a
record that includes deposits from events such as major
floods, which are too rare or violent to be observed
adequately in modern situations, but yet may form a significant part of the sedimentary record. Conversely, the
features observed in a river at low stage, when it is most
accessible, may be destroyed during flood and never be
preserved in ancient rocks.

This paper would be incomplete without a discussion
of channel systems, and a review of recent progress in the
sedimentology and paleohydrology of meandering river
deposits. However, the author's studies have dealt ex-
clusively with braided alluvium, both ancient and modern,
which will therefore receive greater emphasis. Further
details of some of the ancient braided examples discussed
here will be given in other publications.

THE CLASSIFICATION OF ALLUVIAL CHANNEL SYSTEMS

Schumm (1972 and other papers) classified stable alluvial
channels on the basis of the mode of sediment transport and
the lithology of the channel alluvium. His use of the
term stable refers to lack of change in the vertical
dimension; that is, a stable channel is one subject to no
progressive aggradation or degradation. Stability in
horizontal dimensions (ie. on the alluvial surface) is
another aspect of channel function which, to avoid confusion,
I will term <u>lateral stability</u>, in contrast to the <u>vertical
stability</u> of Schumm's definition. Lateral stability is an
indication of the ease or otherwise with which a channel
can relocate on the alluvial surface by lateral or downslope migration. It has a very important influence on the
long-term sedimentary record, as explained in later
sections of this paper. It is reflected in Schumm's
classification, although not developed by him. Thus
suspended-load channels transport a large part of their
total load in suspension; for this reason they have a high
proportion of silt and clay in their banks, which are
therefore relatively resistant to lateral erosion. Conversely, bedload channels transport relatively little suspended sediment, and have easily-eroded channel banks containing little silt or clay.

Another aspect of Schumm's classification is his distinction between single and multiple channels, but he is
at variance with most other authors in placing braided
rivers in the single-channel category: "Braided channels

are single-channel bedload rivers which at low water have islands of sediment or relatively permanent vegetated islands ..." (Schumm 1968b, p 1579). It is hard to see the logic in regarding a braided system as single-channelled because it occasionally floods to occupy one continuous water course, unless it can be shown (and I am not aware of any example) that under these conditions the braid bars, islands and channels are reworked into an entirely different system. Following Schumm's reasoning, one could term meandering channels straight, because they become so (or at least of low sinuosity) when they flood their meander belts. A more logical approach is to classify channels according to their condition at a commonly-occurring medium stage, which need not be defined rigorously, but should be below bankfull. This approach also avoids the problem that at very low stage some single channels diverge around larger bedforms such as dunes, but should not then be regarded as braided.

Most authors have adopted the morphological channel classification of Leopold and Wolman (1957), who recognised straight, braided and meandering categories. Straight channels are rare, except in cases of bedrock control, and will not be considered further here. Leopold and Wolman (p 53) defined a braided river as "... one which flows in two or more anastomosing channels around alluvial islands". They regarded a single channel division around a bar or island as constituting braiding, a view which Allen (1965, p 96) and other authors have accepted. Leopold and Wolman (1957) defined meandering rivers as those with channel sinuosity greater than 1.5. Thus different criteria are used for recognition of the two most important channel categories: sinuosity defines the meandering type, while channel multiplicity defines braided systems.

A better way to define the two major types is to use one single criterion, namely the distinction between single-channel and multi-channel systems. To make this dis-tinction quantitatively, a 'braid' is defined as a single island or bar enclosed by one channel division and one junction, and the number of braids per mean meander wave-length is defined as the braiding parameter. There is usually no difficulty in determining mean "meander" wave-length for a braided reach, because although low, the mean sinuosity for the reach is invariably greater than 1.0. A multi-channel system is then defined as having a braiding parameter less than one. Using this definition, the dis-tinction between single- and multi-channel systems is based on the average characteristics of a reach, and is not in-fluenced by the presents of a few isolated islands (Figure 1). This avoids what seems to be an illogicality in Leopold and Wolman's classification, namely that one island in an otherwise meandering single-channel reach should constitute a braiding situation. Another advantage of this proposal is that it highlights the distinction between relatively stable (single-channel) and laterally unstable (multi-channel) systems, although there are all gradations between the two extremes, as indicated by

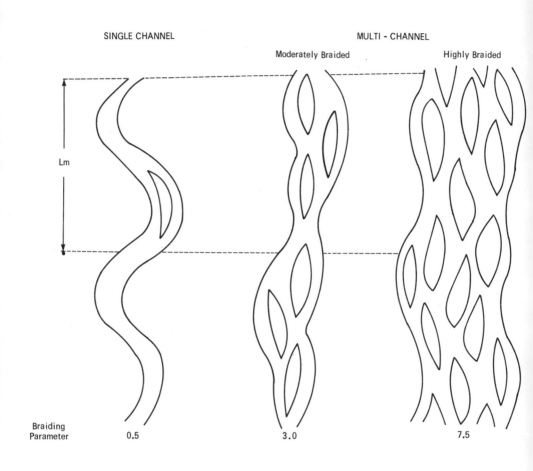

<u>Figure 1</u>   Channel patterns defined by braiding parameter
(see text).   Lm = mean meander wavelength

Schumm (1972).   It also allows one to make arbitrary dis-
tinctions between different degrees of braiding: I suggest
a moderately braided reach is one with a braiding parameter
between 1 and 4, while a highly braided reach has a braiding
parameter greater than 4 (Figure 1).   The braiding index of
Brice (1964) has a similar function, but it gives an
impression of numerical accuracy that is misleading, because
it depends on measurements which vary with river stage
(Rust 1972a, pp 223-225).   This is not so with the braiding
parameter, for unless the river system is at or near complete inundation, the number of braids in a given reach can
still be determined.

Channel types can be further subdivided on the basis
of sinuosity, which can also be quantified as a mean value
for a given reach.   Various authors have chosen different
values for the boundary between low and high sinuosity
channels: 1.3 (Moody-Stuart 1966), 1.5 (Leopold and
Wolman 1957), and 1.7 (Leeder 1973b).   My own choice is for
the median value, 1.5, which gives the four channel categories of Table 1.   Two of these are by far the most
abundant: low sinuosity multi-channel systems (commonly
termed braided), and high sinuosity single-channel systems,
generally described as meandering.   Low sinuosity (or
straight) single-channel systems are rare in nature, although
Schumm (1968a) recognised fairly extensive examples in
"prior-stream" paleochannels of the Murrumbidgee River,
Australia.   Moody-Stuart (1966) claimed recognition of
low sinuosity channel deposits in Devonian rocks on Spits-
bergen, and Leeder (1973a) similarly interpreted coarse
units in the Scottish Devonian, which lack upward fining
and consistent upward change in type and scale of sedimentary structures.   These features could be explained by
the interruption and superimposition of successive point bar
units, and the case for distinctive low sinuosity channel
deposits is at best uncertain.

Multi-channel systems of high sinuosity are also rare,
but they certainly do exist.   They develop on low slopes,
where anabranches are stabilised by vegetation, and are an
exception to the general rule that multi-channel patterns
indicate relative bank instability.   The term   anastomosing
is used in Australia for this type of system (Schumm 1968b,
p 1580), and has been adopted in Canada by Smith (1973).   A
separate term is needed;   anastomosing seems to be the
most suitable, although given as a synonym for braided in
the AGI glossary (1972).

HIGH SINUOSITY SINGLE-CHANNEL ALLUVIAL DEPOSITS

The modern processes of ancient successions resulting from
the activity of meandering channels have been discussed by
many authors, including Allen (1965), Reineck and Singh
(1975, pp 225-253), and Walker (1976).   Walker showed that
the depositional model for meandering fluvial systems is
well enough established to provide a standard against which
variations can be assessed.   The model (Figures 2 and 16,
right column) has two main components: coarse channel units
formed of sand carried by tractive transport and deposited

4. Ancient alluvial successions

Table 1: Classification of channel patterns

	Single-channel (Braiding Parameter <1)	Multi-channel (B.P. >1)
Low sinuosity (<1.5)	Straight	BRAIDED
High sinuosity (>1.5)	MEANDERING	Anastomosing

by lateral accretion on point bars, and flood basin units
formed by vertical accretion of sediment from suspension in
flood waters. The channel and flood basin units alternate,
together comprising fining-upward cycles. Channel units
commonly start with an erosional base covered by lag gravel,
passing up to sand, which decreases in grain size upwards.
The magnitude of sedimentary structures (chiefly cosets of
trough cross-strata) also decreases upwards; sets of horizontally-stratified sand may be interspersed.   Each
cycle can be regarded as formed by more or less continuous
lateral migration of a channel, followed by overbank vertical accretion on levees and in the flood basin. However,
Miall (1977) pointed out that some cycles are more complex
then the model predicts, and the thickness of others (more
than 20 m) implies much larger rivers than is suggested by
the scale of their sedimentary structures. Most of these
features can be explained by superimposition of channel
units without the intervention of overbank deposits.

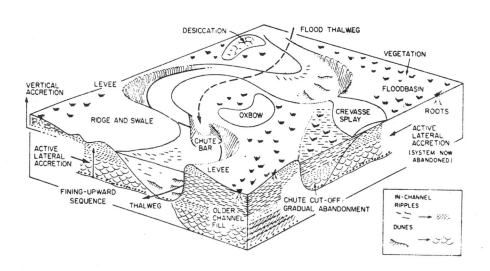

Figure 2   Depositional model for meandering channel
system (from Walker 1976, Figure 1)

The start of a new cycle results from meander cut-off, either by flow through a swale on a point bar surface (chute cut-off) or by breaching of the neck of a meander (neck cut-off), or by avulsion (Walker 1976, Figures 4 and 8A). The degree of change in the river course increases in the order described above, as does the length of channel abandoned. Abandoned channels are gradually filled by vertical accretion deposits during floods, to form plugs of silt and clay which are relatively resistant to subsequent lateral erosion, and therefore exert an important degree of control on the location of the meander belt.

Variation from this model mostly consist of differences in the ratio of lateral to vertical deposits, and variation in the nature of the channel unit. The latter type of variability was discussed by Allen (1964, 1970) who related it to variation in channel sinuosity and stream power. The proportion of flat-bedded sandstone in the channel unit was shown to increase with decreasing sinuosity, which in turn relates to greater source proximity. An increased proportion of cross-bedded as opposed to cross-laminated sandstone (essentially dune versus ripple cross-stratification) indicates an increase in stream power, due to greater channel depth, or slope, or both (Allen 1970, p 145).

Variation in the ratio of vertical to lateral deposits reflects the proportion of suspended as opposed to bedload transported, a basic feature of Schumm's classification, discussed earlier. This ratio is in turn a function of source proximity, for the proportion of bedload tends to decrease with distance travelled. In ancient rocks, however, it is difficult to separate this effect from that due to lithological variation in source material, and erosion may locally reduce either coarse or fine unit thickness. Vegetation is another important factor which increases the proportion of overbank fines in a self-perpetuating manner. Plants preferentially colonise the most stable banks, which are formed of fine-grained, cohesive sediment, and the plants in turn trap more fine sediment during overbank floods. Clearly the appearance of widespread terrestrial vegetation with the evolution of vascular plants during the Upper Paleozoic was an important development in fluvial paleoecology. Schumm (1968b) speculated that bedload (or braided) channels were ubiquitous before this time – a reasonable idea which merits further investigation. Also worth considering is the effect of the evolution of flowering plants, particularly grasses, in the Cenozoic.

The Endrick River, Scotland, is a meandering stream with relatively coarse bed sediment that includes gravel (Bluck 1971). In a personal communication (1972) Bluck stated that the gravel bars deposit pebbly sand at high stage, which is partly reworked at low stage into isolated layers of framework gravel. Fining-upward cycles are common, with overbank silt and clay comprising one to two thirds of each cycle (Bluck 1971, Figures 21, 22 and 23). The Endrick deposits thus depart from the meandering fluvial model only in the relative thickness of the

gravelly part of the channel unit, and the presence of
scattered pebbles higher in each cycle. Similar pebbly
sand sequences were reported by Jackson (1976) in the lower
Wabash River. He found that facies sequences differ in
relation to velocity zones around meanders: the "fully
developed" facies shows regular upward fining, and about
one third overbank mud, whereas pebbles in the "transitional" and "intermediate" facies persist to mid-sequence.
Preservation of facies sequences depends on curvature of
the meander, and its mode of formation (Jackson 1976,
Figures 18 and 19).

Coarse-grained point bar deposits in the meandering
Amite River, Louisiana, depart considerably from the
established model (McGowen and Garner 1970). Their description is based mainly on trenched sections of one point bar,
supplemented by surface observations of other bars in the
Amite, and in the Colorado River, Texas. Unusual features
are the lack of fining-upward sequences, and the apparent
rarity of fine overbank sediment. According to McGowen and
Garner (p 81) the rivers would probably have braided
patterns if their banks were not densely vegetated.
However, braided rivers in diverse climatic environments
flow through densely vegetated regions, for example the
proglacial Donjek (Rust 1972a) and the tropical Brahmaputra
(Coleman 1969). These rivers have unstable banks because
they lack cohesive overbank deposits, not because of inhi-
bited plant growth. The question therefore is whether
overbank fines are more extensive elsewhere on the Amite
floodplain than in the sequences reported by McGowen and
Garner.

The Amite and Colorado bed materials were described as
coarse sand and pebble to cobble gravel, but this appears
to be a field description biased towards coarse particle
content, because the size data given (McGowen and Garner,
Figures 22 and 23) show most samples with mean sizes in
the medium to coarse sand range. For example, a
"homogeneous gravel" has mean size 0.7 $\emptyset$, standard deviation
0.7, and zero skewness; with these statistics it can have
little more than a few percent of the finest gravel-sized
particles. In general it appears that the Amite and
Colorado bed sediment resembles that of the Endrick (Bluck
1971): sand and pebbly sand, with pebbles locally concentrated in lag veneers. The lack of fining-upward sequences
is due to the introduction of coarse material into the
upper part of the point bar unit by the migration of chute
bars during flood. This is attributed to the "flashy"
nature of discharge variation in the Amite: "Most point-
bar accretion is accomplished during extreme flood when
the channel tends to straighten its course through a shift
of the thread of maximum surface velocity toward the convex
bank" (McGowen and Garner 1970, p 91). In meandering
rivers with broader discharge fluctuations flood flow does
not carry the coarsest load over the point bars, or if it
does, chute cut-off and abandonment of the original channel
is likely to result. The point bar sequence of the Amite
River must therefore be regarded as an unusual type of
uncertain applicability elsewhere.

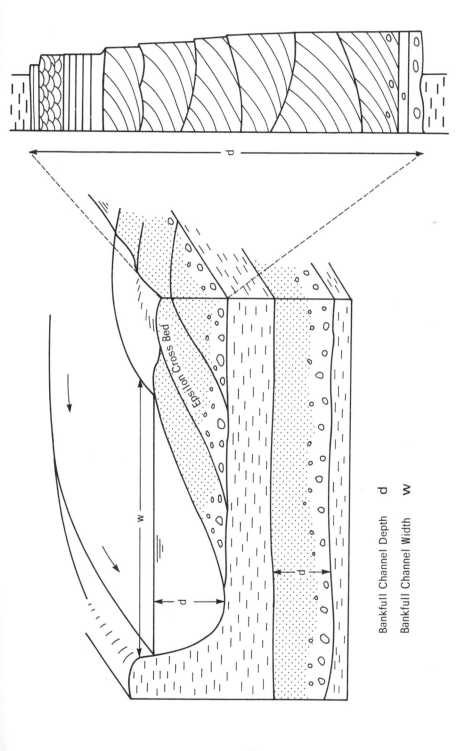

Bankfull Channel Depth   d

Bankfull Channel Width   w

Figure 3   Paleochannel depth for meabdering systems (after Leeder 1963b, Figure 1)

4. Ancient alluvial successions

Several authors have attempted to reconstruct dimensions and hydrologic parameters of paleochannels based on structures in ancient rocks, and modern fluvial data. This procedure is feasible for meandering systems, because, as discussed above, the controlling factors are quite well understood. The bankfull depth of the paleochannel (d) can be determined from the thickness of the channel unit, or from the thickness of epsilon cross-beds, if present (Figure 3). Epsilon cross-beds (Allen 1963) represent the ancient depositional surfaces of point bars, but they are commonly obscured by down-channel migration of smaller-scale bedforms. If they are present, bankfull channel width (w) can be determined as 3/2 x epsilon cross-stratal width; if not, the following relationship can be used:

$$\log w = 1.54 \log d + 0.83$$

or $w = 6.8d^{1.54}$, in metric units (other equations given in English units). This relationship is based on channels with sinuosity greater than 1.7; it has a correlation coefficient of 0.91, and standard deviation of 0.35 log units (Leeder 1973b). However, it should be emphasized that the basic assumption is that paleochannel depth equals the thickness of the coarse unit. If successive coarse units are deposited without intervening overbank fines, a serious overestimation of paleochannel depth results. Underestimation occurs if the top of a channel unit is eroded and replaced by overbank fines, but this is rare because erosion is accomplished by lateral shifting and is normally followed by deposition of another channel unit.

Other paleohydrologic parameters can be determined from equations based on a variety of rivers:

$$Lm = 10.9w^{1.01} \qquad \text{(Leopold, Wolman \& Miller 1964)}$$

$$Lm = 106\bar{Q}^{0.46} \qquad \text{(Carlson 1965)}$$

$$s = \frac{30 \ (w/d)^{0.95}}{w^{0.98}} \qquad \text{(Schumm 1972)}$$

$$M = \frac{(Sc \ x \ w) + (Sb \ x \ 2d)}{w + 2d} \qquad \text{(Schumm 1968a)}$$

$$P = 0.94 \ M^{-0.25} \qquad \text{(Schumm 1963)}$$

Where Lm = meander wavelength, $\bar{Q}$ = mean annual discharge, s = slope, M is a channel lithology parameter, Sc is the % silt + clay in the channel alluvium, Sb is the same in the bank, and P is channel sinuosity. In most cases the authors quote confidence limits, and the equations will generally give order-of-magnitude estimates, but each derivation involves additional assumptions which add to the margin of error.

Another source of uncertainty is the fact that Schumm's equations were derived for vertically stable channels in sub-humid or semi-arid climates, and exclude gravels (Schumm 1972, p 100). It is obvious that ancient alluvium must have progressively aggraded, or it would not be preserved. The presence of concretionary carbonates in over-- bank units is commonly taken to be an indication of relative aridity, but other factors may be effective, such

as low sedimentation rates (Leeder 1975). Thus paleo-
climates cannot be diagnosed precisely, and the errors involved in these departures from Schumm's conditions are
unknown.

In some instances preservation of ancient alluvial deposits may allow recognition of point bar ridges and swales
(Nami 1976; Padgett and Ehrlich 1976). In these cases
meander wavelength and radius can be measured directly, and
other parameters such as channel width, slope, and mean
annual discharge estimated therefrom.

LOW SINUOSITY MULTI-CHANNEL DEPOSITS

Introduction

In comparison with meandering systems, the processes and
products of low sinuosity multi-channel (braided) rivers
are poorly understood. This is partly due to their
inherent variability, and partly results from neglect. For
example, Visher's (1972) review ignored braided rivers,
while Leopold, Wolman and Miller (1964) and Allen (1965)
emphasized meandering systems. Thus, as Walker (1976)
pointed out, we have no depositional model for braided
systems with the predictive capability of the meandering
model described above. However, there has been increased
interest in braided rivers in recent years, as reviewed by
Miall (1977), who proposed four sedimentary models based on
facies assemblages and vertical sequences. His facies
types will be used, where possible, in this paper.

Braided alluvial deposits can be divided into three
types based on their dominant lithology: gravel, sand and
silt. Silt-dominated braided rivers are rare, and have
received little attention; the only detailed account is
that of the Yellow River (Chien 1961). In many cases there
is a downstream transition from one lithotype to another,
for example that from gravel- to sand-dominated reaches in
the Platte Rivers (Smith 1970, 1971). The Slims River,
Yukon, shows transitional changes from gravel- to sand- to
silt-dominated reaches within its 14-mile length
(Fahnestock 1970).

The distinction made here between gravel- and sand-
dominated systems is that the principal lithotype of the
former is framework, or clast-supported gravel, with no
upper size limit. Gravel-sized particles may be present in
sandy deposits, but if so they are matrix-supported except
where reworked as lag concentrates. In effect this criterion distinguishes between rivers which in flood transport
gravel as bedload and sand in suspension, from those which
move a mixture of sand and pebbles along the bed. In the
former case gravel is deposited first during falling stage,
and is infiltrated later by sand except for rare situations
in which the gravel is preserved in an openwork condition.
Matrix-supported gravel, or pebbly sand results from simultaneous deposition of sand and pebbles, which cannot be
much in excess of sand size.

The same criterion can be used to distinguish gravelly
braided deposits from those of coarse-grained meandering
rivers such as the Endrick and Amite, discussed above. Unfortunately the definition proposed here is not always

applicable to descriptions in the literature because some
authors have reported the presence of gravel or conglomerate
without stating whether or not it is of framework type, and
without sufficient grain size data to make reasonable
inference. Pettijohn (1975, p 154) noted the tendency to
use the term gravel or conglomerate for sediments with
relatively minor proportions of gravel-sized clasts.

Smith (1970) and others have observed that longitudinal
bars, elongate parallel to flow are typical of gravel
reaches, whereas transverse bars, with slip faces normal to
local flow direction are prevalent in sandy reaches. A
complex morphological nomenclature for bars has arisen over
the years (Miall 1977) but for application to ancient sediments bar classification should be simple, and must incorporate internal structures (Rust 1975, p 238; Hein and
Walker 1977, p 565). Hence I propose the following:
Longitudinal bars: elongate parallel to the general flow
direction; internal structure dominated by horizontally
stratified, imbricate, framework gravel; characteristic of
proximal reaches.
Transverse bars: various shapes, with depositional slip
faces transverse to local flow directions at downstream
(commonly lateral) margins; internal structure predominantly planar cross-stratified sand or pebbly sand;
characteristic of distal reaches.

Other types described in the literature can be regarded
as sub-types or modifications of the above: linguoid bars
(Collison 1970) are staggered transverse bars; diagonal
bars (Church and Gilbert 1975, p 58) are asymmetric longitudinal bars (Miall 1977); point and side bars form at
channel margins by coalescence of other bar types *(op cit)*.
An important but rare bar type which does not fit into this
classification is the giant braid bar of Pleistocene catastrophic floods (Bretz and others 1956; Malde 1968). These
bars are elongate parallel to the flow, and contain
bouldery framework gravel in giant sets of planar cross-
strata. Their affinities are obviously towards longitudinal
bars, and it is proposed here that they should be classified as giant longitudinal bars, with the following
characteristics: elongate parallel to flow; internal
structure dominated by giant sets of planar cross-beds in
boulder gravel; characteristic of deep, catastrophic floods.

Gravel-dominated braided deposits

(i)   Bar formation and migration

Gravel bars are basic structures not only in proximal
reaches, but for considerable distances downstream. For
example, gravel (mostly framework type) is the principal
lithotype in the Donjek outwash 50 km from its source
(Williams and Rust 1969, p 652). Even at this distance
finer sediments accumulate as relatively minor bodies peri-
pheral to gravel bars, except in inactive, well-vegetated,
areas.

Fundamental to the interpretation of structures in
proximal fluvial gravels is an understanding of the
formation and migration of longitudinal gravel bars (Hein
and Walker 1977, p 563). The commonly accepted model is
that of Leopold and Wolman (1957) based on observations in

a flume and of individual braid bars in a small gravelly stream . Leopold and Wolman proposed that longitudinal bars are initiated by deposition of the coarser fractions of the bedload in mid-channel. Subsequent growth takes place by addition of finer sediment on top of and downstream of the bar nucleus, which may be raised above water, and later vegetated. However, it is by no means certain that this model of bar growth is applicable to large gravelly braided systems, although several authors have attempted to do so (Ore 1964, p 3; Costello and Walker 1972, p 397). The major problem is that direct observation is impossible during flood stage, when major bar changes are to be expected.

Smith (1974) and Hein and Walker (1977) attempted to overcome this problem by studying the effects of diurnal rise of discharge on bar formation and migration in the Kicking Horse outwash, British Columbia. The flow conditions usually permitted observation of the bed through the water and channels were wadable. Smith termed active bars with predominantly depositional morphology "unit bars" and observed their growth downstream or laterally. Hein and Walker (1977) postulated an initial stage of bar formation from "diffuse gravel sheets", which develop into longitudinal or diagonal bars with horizontal stratification, or transverse bars with cross-strata. These observations are of intrinsic interest, but there is some doubt that they are a suitable model for processes occurring during major flood cycles,which probably form the greater part of deposits preserved in ancient successions. In floods, however, direct observation and wading are impossible, and our only recourse is to deduction based on the structures preserved in braided alluvium.

The dominant structure in proximal outwash gravels is poorly defined horizontal bedding (McDonald and Banerjee 1971, p 1297; Church and Gilbert 1975, p 61 and Figure 34; Rust 1975, pp 240-241). Its internal fabric, and that of massive gravel in the same deposits, is consistent with deposition on subhorizontal surfaces (Rust 1975, p 245). The same combination of structure and fabric is prevalent in other types of coarse braided alluvium (Ore 1964, p 9; Smith 1970, p 2999). However, if Leopold and Wolman's (1957) model were valid for flood stage processes in rivers of this sort, one would expect cross-bedded gravel to predominate, whether formed by migration of riffles as low angle cross-strata, or slip faces (high angle) as observed by Smith (1974, p 219) in diurnally-induced small-scale bar movements.

Another implication of Leopold and Wolman's model is that if longitudinal gravel bars form by coarse fraction deposition during falling stage, the bars should be in a state of disequilibrium at stages high enough to transport all available material, and a flat bed would result if this stage were maintained long enough. That this is not so is indicated by the giant longitudinal bars of the Melon Gravel (Malde 1968) and other deposits formed by catastrophic floods (Bretz and others 1956). These bars are so large that they must have been primary structures, only superficially modified during falling stage. It is

reasonable to assume that a similar condition prevails on
proximal braided alluvium during flood stage, ie. the
longitudinal gravel bars are equilibrium forms when all
fractions of the bed material are in motion. Thus the
braid bars are maintained during flood, and may be analogous
to antidunes on a sand bed - indeed it is likely that upper
regime conditions prevail when they are active. Probably
they migrate upstream or downstream (accumulating gravel on
very gently sloping surfaces), or accrete vertically, depending on the circumstances (Rust 1975, pp 246-7). The
resulting deposit is therefore characterised by poorly
defined subhorizontal stratification. This condition is
maintained during high flood stage in proximal braided
alluvium, because the ratio of water depth to mean particle
diameter commonly remains too low to allow the formation of
slip-faces and the generation of cross-stratal units. This
was not so in the case of the Melon-Gravel, for which Malde
(1968) estimated water depths of several hundred feet; hence
the giant longitudinal bars developed slip faces and
generated giant sets of cross-strata.

(ii) High slope proximal deposits (alluvial fans)

The distinction between high and low slope proximal environments is somewhat arbitrary. For example, the main upper
tributaries of the Donjek River, Yukon, are glacier-fed
braided streams which have built alluvial fans where they
enter the main valley (Rust 1972a, Figure 2A). This is a
response to the recently-exposed glacial relief in which
tributary valleys join the trunk valley at elevations considerably above its floor. The slopes of the fans vary
inversely with size, the largest (Steele Creek fan) having
a slope of 0.013 compared with 0.006 for the Donjek River
in this, its most proximal reach. A similar range of slopes
is encompassed within a single outwash system in the Scott
fan, Alaska, as termed by Boothroyd and Ashley (1975). A
twofold difference in gradient is readily apparent in modern
landforms, but would be of little or no help in diagnosing
ancient environments. In addition, the stratal type most
common in alluvial fans is formed by stream flow, and
strongly resembles that of proximal braided rivers; imbricate,
coarse, framework gravel (Rust 1972a, Figure 3; 1975,
Figure 4), with poorly defined horizontal strata in sets
bounded by indistinct low-angle erosion surfaces. This is
equivalent to facies G_m of Miall (1977) and facies 6 of
Rust (1972a). The distinction made by Bull (1972) between
sheetflood and stream channel sediments on alluvial fans is
thought to have doubtful application to the identification
of ancient fan and river deposits.

A more useful distinction is that minor sets of cross-
bedded gravel, and horizontal or cross-stratified sand are
relatively more abundant in the fan deposits than in the
proximal Donjek outwash. This accords with Bull's (1972,
p 73) observation that lithological variation is a distinctive feature of fan deposits, which is also reflected
in rapid textural changes downfan. Decreased mean grain
size is the most obvious trend, but sorting also tends to
increase downfan, as does clast roundness (Blissenbach
1954). Changes of this sort require distances on the order
of tens of kilometers in rivers, but can lead to fine valley

flat deposits in a few kilometers of downfan transition.

Another distinction is that debris flow deposits are common on proximal parts of fans, where they alternate with stream flow deposits in vertical section. The debris flow deposits commonly lack stratification and imbrication, and are poorly sorted, with megaclasts (commonly angular) supported by a matrix which may contain significant amounts of mud-sized material. The frequency of debris flows on alluvial fans is a function of proximity to high relief, and of rainfall, vegetation, and source lithology (Bull 1972, p 69). The low water table causes rapid infiltration which arrests most debris flows on the proximal parts of the fan (this phenomenon also influences stream flow). A few large debris flows may reach the trunk river where they are likely to be reworked, whereas their chances of preservation on the fan are relatively high. Thus the major causes of difference between fan and proximal river deposits are slope and infiltration capacity, which give rise to greater heterogeneity on the fan. There is no reason why these physical differences should not be similar in proglacial and semi-arid environments. However, the presence of evaporites and concretionary soil horizons in fans is an indication of relative aridity (Bull 1972, Figure 2).

The Cannes de Roche Formation is a Mississippian deposit in eastern Gaspé (Figure 4), with three distinct members (Alcock 1935; McGerrigle 1950). The Lower Member is composed chiefly of breccias, and is interpreted as a proximal alluvial fan deposit, transitional upward and laterally into distal fan and valley flat deposits of the Middle Member (Rust 1977b). Both members are red, the intensity of colouration increasing with decreasing grain size.

The Lower Member locally contains very coarse breccias which are interpreted as debris flow deposits. Bedding is rarely apparent, except where it is revealed by impersistent lenses of sandstone (Figure 5). The clasts are sharply angular, range up to 1 m in size, and are commonly supported by a matrix of muddy sandstone, although the proportion of matrix rarely exceeds 50%. Clast orientation is highly variable: it is commonly random, but in some of the coarser deposits tabular particles tend to be subhorizontal (Figure 5). In finer, presumably more distal deposits, flat particles tend to lie at high angles to the bedding, with random distribution about a subvertical axis. A similar fabric was observed by the author in distal lobes of debris flow deposits in the upper Donjek Valley. Bull (1972, pp 70-71) interpreted horizontal as opposed to vertical clast orientation as a function of relatively low and high flow viscosity respectively. Johnson (1970, pp 433-434) pointed out that to explain the characteristically steep terminations of debris flows, they must have finite strength as well as viscous properties. This suggests that the subvertical clast orientation in the finer Cannes de Roche debris flow deposits is due to the increased viscosity and strength of distal lobes resulting from water loss by infiltration into the fan.

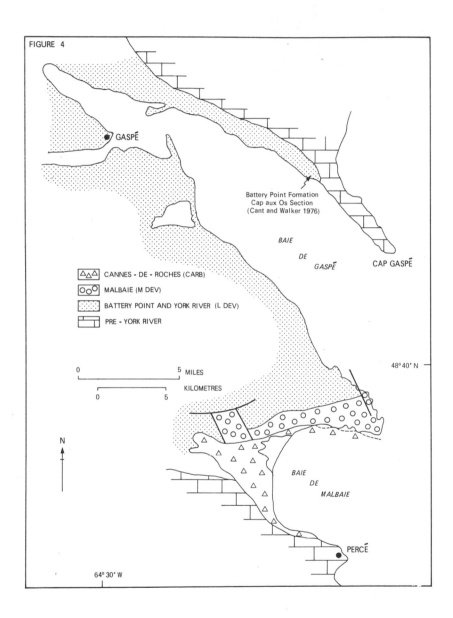

Figure 4   Geological map of eastern Gaspé
Peninsula, Quebec, showing
localities and formations discussed
in text

Figure 5   Debris flow deposit, Lower Member, Cannes
de Roche Formation.   Tape 20 cm

Figure 6   Concretionary limestone in upper part of
a fining-upward unit, Lower Member,
Cannes de Roche Formation.   Tape 50 cm

4. Ancient alluvial successions

The variability of fabric (or lack of it) described
above for debris flow deposits is in marked contrast to
the relatively uniform fabric of stream flow deposits:
upstream dip of tabular clasts at moderate angles, commonly
20-35° (Rust 1975). Stream flow deposits predominate in
the finer parts of the Lower Member succession, and are
mainly characterised by horizontal strata, with minor
planar and trough cross-stratified units. The principal
lithology is fine breccia, with mean size in the 1 to 2 cm
range, and very angular to subangular particles. The
angularity of clasts in relatively distal deposits is
unusual, for Blissenbach (1954) reported a change in
roundness from 0.1 at the apex of a modern fan to 0.7 at
its base 4 miles away. The principal source rock for the
Cannes de Roche breccias is a microcrystalline limestone,
which evidently has a high fracture susceptibility.

The breccias occur in approximately tabular units
0.5-3 m thick which grade up into or alternate with thinner
units of silty sandstone. The latter units could be the
fine remnants of debris flows, but it is thought more
likely that they are the deposits of finer stream flow
loads, which settled on distal parts of the fan as flood
waters infiltrated. The upper parts of these fine units are
the principal location of concretionary deposits of calcium
carbonate and minor microcrystalline silica, thought to be
soil calcretes and silcretes (Figure 6). The breccia/
sandstone succession is cut in places by channels, most of
which are wide and shallow, but a few have steep lateral
margins (Figure 7).

The Middle Member of the Cannes de Roche Formation is
interpreted as a valley flat deposit. It is characterised
by faintly laminated to apparently massive red mudstone
and siltstone, with minor interbeds of fine breccia and
horizons of concretionary limestone (Figure 8). The
breccias are identical to those of the Lower Member, except
that they are finer grained and are almost devoid of cross-
strata. They are almost entirely composed of thin horizontal strata, and are interpreted as the distal equivalents
of the larger stream and/or debris flow deposits of the
Lower Member. The similarity of the breccia units indicates
that the two members are lateral equivalents, which is also
predicted by Walther's Law, for the members are completely
transitional in stratigraphic succession.

(iii) Low slope proximal deposits

The proximal outwash of the Donjek River is a modern example
of this type of deposit (Rust 1972a, 1975). It closely
resembles the Scott outwash (Boothroyd and Ashley 1975),
and is typical of the Scott type depositional model of
Miall (1977). The basic stratification-forming process is
related to bar genesis, as discussed in an earlier section.
Ancient equivalents are found in the conglomerate units of
the Malbaie Formation, a Middle Devonian terrestrial
succession in eastern Gaspé (McGerrigle 1950, McGregor
1973). The formation consists of abruptly alternated units
of sandstone and conglomerate 10 to 200 m thick (Rust 1976,
Figure 1; 1977a). The coarse units consist of medium to

Figure 7   Fine breccia filling steep-sided
            channel, Lower Member, Cannes de
            Roche Formation

coarse framework conglomerate (maximum clast size 10 to
40 cm), of which 80% is horizontally bedded, and 20%
planar cross-bedded (Figure 9); (facies Gm and Gp of Miall,
1977).   Sandstone (mostly Miall's Sp and Sh) forms about
3% of the conglomerate successions, although some conglomerate cross-beds are locally sandy, to the extent of being
matrix-supported.   Pebble orientation is well developed in
the horizontally-bedded conglomerate, with ab planes
dipping upstream and a axes perpendicular to paleoflow
directions, a similar fabric to that observed on and within
longitudinal bars and shallow channels of proximal braided
outwash (Rust 1972b; 1975).

    Paleocurrent directions determined from conglomerate
fabric in horizontal strata are remarkably consistent over
the whole area (Figure 10), and give high vector magnitudes.
The relatively low directional variance of gravel fabric
was also noted in recent outwash by Bluck (1974), and is
due to formation at high stage, when the flow transports
gravel over the bars and in channels with minimum
deflection from the downslope direction.   Paleocurrents
derived from conglomerate cross-beds are much more variable,
but yield a similar overall mean vector.   This could be due
to formation of conglomerate cross-beds by migration of
independent transverse bars, but is considered unlikely.
The cross-bed sets are relatively small (mean volume 125 m^3),
commonly show lateral transition to horizontally-bedded
conglomerate, and make up only 20% of the conglomerate

Figure 8   Part of Middle Member succession, Cannes
de Roche Formation, showing mudstone with
thin breccia units and a concretionary
limestone horizon.   Tape 50 cm

Figure 9   Horizontal bedding and planar cross-
bedding in a Malbaie Formation
conglomerate unit.

succession. It is therefore concluded that they form as
lateral modifications on longitudinal bars during falling
stage, when flow diverges away from emerging bar axes
(Figure 11).

Maill (1977, Figure 10) summarised cycles in recent
braided deposits, of which the Scott type (his Figure 12)
is the most appropriate to low slope proximal gravels. It
comprises alternations of massive to horizontally bedded
gravel (Facies Gm) and various sand facies (St, Sh, Sr)
interpreted as formed under progressively decreasing
energy levels during a flood cycle. Sand beds of similar
type were observed on the proximal Donjek outwash (Rust
1972a), but seldom survive the following flood, as indicated
by their rarity (less than 1%) in vertical sections.
Evidently most flood cycles in proximal deposits are interrupted, so that the gravel units commonly occur in erosional
contact with the underlying gravel. The conglomerates of
the Malbaie Formation record a similar prevalence of
interrupted flood cycles.

(iv) Distal gravel deposits

Gravel persists as the dominant lithotype for considerable
distances downstream in several braided rivers, for example
the South Platte (Smith 1970) and the Donjek (Williams &
Rust 1969). Area 2 of the Donjek (Rust 1972a) is 50 km
from the river's glacial source, and is characterised by an
active tract dominated by framework gravel, and inactive

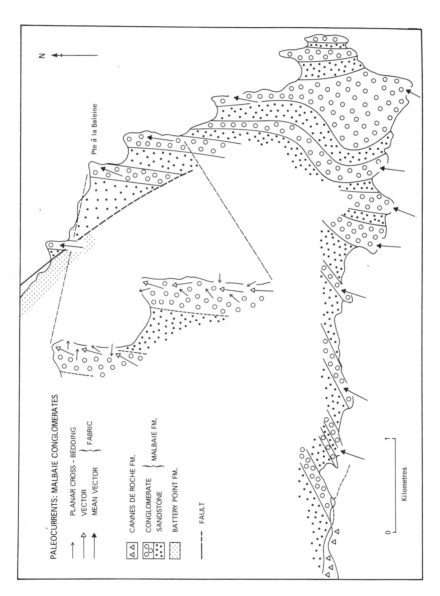

Figure 10 Paleocurrents in Malbaie conglomerates

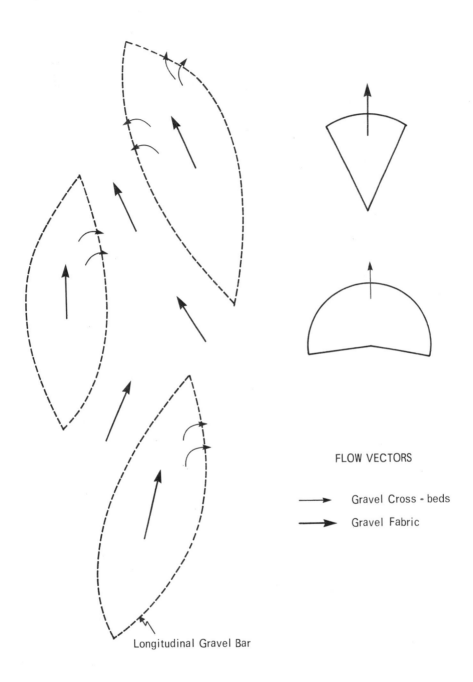

FLOW VECTORS

→ Gravel Cross - beds

➤ Gravel Fabric

Longitudinal Gravel Bar

Figure 11   Bar model to explain differences between
paleocurrents derived from conglomerate
fabric and cross-beds in the Malbaie
Formation

areas on which finer sediment accumulates, eventually
supporting dense vegetation (Levels 1 and 4, with intermediate categories, Williams and Rust 1969). Compared with
proximal reaches, it is likely that the relative depth of
the main active channels and the relatively fine nature of
the gravel would produce bedforms with slip faces during
flood.   Under broadly similar circumstances, echo sounding
by Fahnestock and Bradley (1972, p 241 and Figure 14) showed
dunes on the fine gravel bed of the Knik River, Alaska. The
grain size of the bed material and the dimensions of the
dunes reached maximum values at the highest intensity of
flood investigated. The three dimensional form of the
dunes was not recorded (this would probably require side-
scan sonar, which would not work well in the shallow, highly
turbulent water), but most likely they were sinuous,
generating trough cross-strata (facies Gt of Miall 1977).
Gravel bars with slip faces were observed in the North
Saskatchewan River, Alberta, by Galay and Neill (1967).
Mean bed material size was 5 cm, and the slip faces were
shown to be gently sinuous, crescentic surfaces. In ancient
successions these would probably be identified as planar
cross-beds (Gp of Miall 1977) unless exposures were particularly extensive.

The Upper Member of the Cannes de Roche Formation
exhibits many of the features described above (Rust 1977b).
The most abundant lithotype is fine-grained framework
gravel in large-scale sets of trough cross-strata (Figure
12).   These sets commonly occur in multiple units with
sharply erosional bases, and tend to fine upwards to
smaller sets of trough cross-stratified sandstone and pebbly
sandstone.   Minor sets of horizontally bedded fine framework gravel and horizontally stratified sandstone are also
present, usually stratigraphically above their respective
troughed units. The horizontally or trough cross-stratified
sandstone passes up into silty mudstone with minor troughed
sandstone units (Figure 12). Plant fossils are abundant, as
logs up to 5 m long in the conglomerate, and as fine frag-
ments in the sandstone and siltstone.

On average the multiple conglomerate/sandstone units
make up 85-90% of the succession, the rest being silty mudstone and minor sandstone. This is interpreted as a distal
braided environment like Areas 2 and 3 of the Donjek, with
the conglomerate/sandstone units being the deposits of the
active tract, and the mudstone and vegetation present on inactive areas. The vertical section at any one location
represents gradual aggradation within an active channel complex, with dunes on the beds of the major channel or
channels generating trough sets of gravel. Continued
aggradation in the channel and/or shifting of a bar over a
channel results in deposition of sand trough sets, or horizontally stratified gravel of sand. Eventually the active
tract migrates elsewhere on the floodplain; the area
becomes inactive, and slow sedimentation of mud with
growth of vegetation takes place, while minor channels
still transport sand during flood and generate small sets
of trough cross-strata.

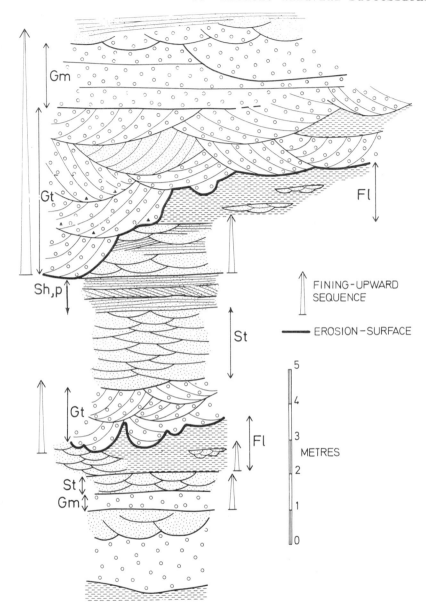

Figure 12   Stratigraphic section of part of the Upper
             Member, Cannes de Roche Formation
             Gt: Trough cross-bedded conglomerate
             Gm: Horizontally bedded conglomerate
             St: Trough cross-stratified sandstone
             Sh,p: Horizontally and planar cross-
                    stratified sandstone
             Fl: Mudstone with minor sandstone trough
                    sets

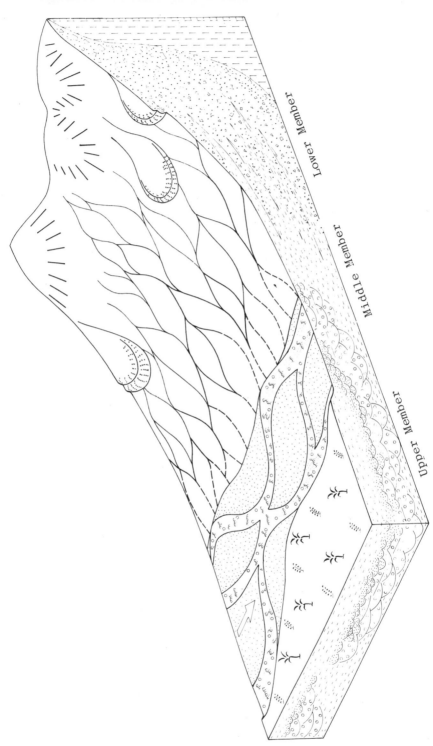

Figure 13 Block diagram for Cannes du Roche depositional environments

Lower Member

Middle Member

Upper Member

The lowermost channel complex of the Cannes de Roche Upper Member is sharply erosional into the Middle Member, but the upward colour gradation from red to grey in the respective mustones links the two members together strati-graphically. Hence, according to Walther's Law, it is reasonable to view the depositional environments of the two members as lateral equivalents (Figure 13). This suggests that the semi-arid redbed/calcrete deposits of the Lower and Middle Members passed laterally and downslope into Upper Member deposits with a very different groundwater regime. Red colouration gave way to grey (siltstone) and buff (conglomerate and sandstone), and plants became abundant. This implies that the Upper Member deposits were the distal part of a braided system draining a more humid region elsehwere. Mean paleocurrent vectors for the lower and Upper Members vary by about 90°, giving support to this interpretation (Rust 1977b).

The deposits described here are broadly similar to the Donjek model of Miall (1977, Figure 13), except that framework conglomerate is much more abundant in the Upper Member of the Cannes de Roche[1], and ripple cross-laminated deposits have not been observed, although they are probably present. The larger proportion of fine-grained deposits in the Cannes de Roche is to be expected. The Donjek River is confined within its glacial valley, whereas most braided deposits preserved in the stratigraphic record were more extensive, or there would be little like-lihood of preservation and subsequent exposure. Hence most ancient systems would have had more space in which inactive areas could develop and the presence of relatively thick fine units and fining-upwards channel sequences should not be considered unusual. Both these features are commonly regarded as typical of meandering fluvial deposits (see earlier section), but the presence of framework gravel as a major lithotype is here considered evidence of a braided system.

Sand-dominated braided systems

(i) Proximal

Sandy braided alluvium will only form in a proximal environ-ment if gravel-sized clasts are not available. Since con-siderably relief and at least occasionally high runoff are implied, these conditions are rarely fulfilled. A modern example is an ephemeral flood deposit formed in Bijou Creek, Colorado (McKee and others 1967), which is dominated by plane-laminated sand (Miall 1977,Figure 15). This structure is formed by upper regime flow, and is analogous to the shallow flood flows which form longitudinal bars in proxi-mal gravel systems. A somewhat different type is the sand-stone of the Malbaie Formation (Rust 1976, 1977a) and similar deposits in the Devonian Peel Sound Formation of Somerset Island, NWT (M.R.Gibling, personal communication).

---

[1] And in Area 2 of the Donjek; however Miall stressed that his model, although named after this river, is based mainly on ancient successions.

4. Ancient alluvial successions

The sandstone units of the Malbaie Formation are
sharply differentiated from its conglomerate units. Extra-
formational gravel-sized clasts are rare, but mudstone
intraclasts (up to 50 cm in longest dimension) are very
abundant. Three major lithological "states" are present
in order of decreasing abundance: erosional surfaces with
mudstone intraclasts; horizontal to low angle ($<10^{o}$)
planar cross-laminated sandstone (LAPL); and trough cross-
stratified sandstone; the latter two having scattered
mudstone intraclasts in some cases (Figures 14, 15). The
erosional surfaces vary from planar, to irregular, to
scoop-shaped, and the intraclastic fill from one clast to
2 m thick. The LAPL sandstone is very similar to the low
angle stratified sandstone (Facies G) of Cant and Walker
(1976, p 110), and commonly has primary current lineation
on the laminae. In cases where the laminae dip, lineations
are found parallel, perpendicular, and oblique to strike.
The term trough cross-stratified sandstone used here includes
isolated structures up to 1.5 m thick, which are probably
scour fills, as well as grouped sets with individual
troughs commonly about 30 cm thick, thought to have been
formed by migrating dunes. It was not always possible to
distinguish between the two types, which were therefore
grouped.

Less common lithotypes, in decreasing order of
abundance are: high angle planar cross-stratified sandstone, ripple-drift cross-laminated silty sandstone to
siltstone, and primary mudstone. The ripple-drift units
commonly show upward fining, and contain *in situ* plant
roots, which, together with the red colouration, are a good
indication of terrestrial deposition. The most
characteristic feature of Malbaie sandstone successions is
their vertical and lateral variability (Figure 15). The
presence of only three major lithotypes, and the vertical
variability renders statistical analysis of the sequence
invalid. The abundance of mudstone intraclasts indicates
that plenty of mud was transported and deposited within
the system, but it rarely survived erosion by later floods.
This implies an ephemeral system which permitted desiccation
of the mud during dry periods. The structures in the sandstone suggest abrupt alternations from upper to lower flow
regime, due to rapid changes of stage, and of depth, as
shallow channels and impersistent bars shifted. These
processes resulted in innumerable erosion surfaces.

In contrast to the Malbaie sandstones, the Bijou
Creek flood deposits were overwhelmingly dominated by
horizontally stratified sand (McKee and others 1967;
Miall 1977, Figure 15). The transport of bridge beams
and concrete slabs indicate that stream competence was
greatly in excess of that required to move sand. The flow
was shallow, of high velocity, and therefore predominantly
in the upper flow regime. Few channels were formed in
the flood, which was evidently of short duration. It is
concluded that sandstones of the Bijou Creek type represent
an even more ephemeral type of alluvial system, which is
likely to be less abundant in the geological record.

Sp

LA
PL

LA
PL

ES

St,

ES

ES

Figure 14   Sedimentary structures in Malbaie sandstones
(Ruler 30 cm): ES: erosional surface with
mudstone intraclasts LAPL: low-angle plane-
laminated sandstone St: trough cross-
stratified sandstone Sp: planar cross-
stratified sandstone

Figure 15   Part of the Malbaie sandstone succession
west of Pte à la Baleine; symbols as in
Figure 14. Plank is 55 cm long

ES

St

ES

LAPL ES Sp ES LAPL ES LAPL

4. Ancient alluvial successions

(ii) Distal

Studies of sand-dominated distal reaches of braided rivers include those on the Brahmaputra (Coleman 1969), the Tana (Collinson 1970), the Platte (Smith 1970), and the South Saskatchewan (Cant 1975; Walker 1976). Miall (1977) used the Platte as a depositional model for this type of braided alluvium.

Walker (1976) and Cant (1977) compared the recent deposits of the South Saskatchewan River with the Lower Devonian Battery Point Formation, in which Cant and Walker (1976) identified repetitive sequences near Cap aux Os, Quebec (Figure 4). The model sequence fines upwards, starting with a scoured surface below a thin layer of mudstone intraclasts, interpreted as a channel-floor lag. This is overlain by poorly-defined large-scale trough cross-beds, and these in turn by interbedded smaller trough cross-strata and planar cross-stratified sandstone. The planar sets commonly show paleocurrent directions about 90° from the trough sets (Figure 16, left column). Interpreted in terms of phenomena observed in the South Saskatchewan, the trough sets form by down-channel migration of sinuous-crested dunes, while the planar cross-stratified sets result from lateral migration of transverse bars. These bars have a complex form, with slip faces commonly migrating at high angles to the channel trend (Walker 1976, Figure 10)). The sequence ends with vertical accretion of relatively minor bar-top deposits, which include small-scale planar cross-strata, rippled siltstones and mudstone.

Other studies of modern sandy braided systems indicate some of the variations which should be expected in ancient deposits, particularly those due to variation in transverse bar shape, and in bedform scale, in turn related to river size. Transverse bars in the Lower Platte River have sinuous to lobate depositional fronts (Smith 1971); those of the Tana are tongue-shaped and arranged en echelon (linguoid bars, Collinson 1970). The Brahmaputra is notable for the magnitude of its bedforms, the largest of which, termed sand waves by Coleman (1969) have heights ranging from 7.5 to 15 m and lengths of 180 to 900 m. They migrate in flood, but as river stage falls they are left in a state of disequilibrium with the flow. This leads to erosion of crests and deposition in troughs, but such large structures persist for considerable periods before complete reworking (Allen 1973). Thus a considerable part of the large-scale structures formed at flood stage will commonly be preserved. Conaghan and Jones (1975) interpreted large-scale structures in the Triassic Hawkesbury Sandstone of New South Wales as formed in a braided river of size comparable to the Brahmaputra.

There is no hard-and-fast distinction between distal braided and meandering fluvial systems; many authors have noted transitions within single rivers (Allen 1965). One should expect a class of ancient fluvial deposit in which it is impossible (and, in fact, relatively unimportant) to determine whether the channel pattern was braided or

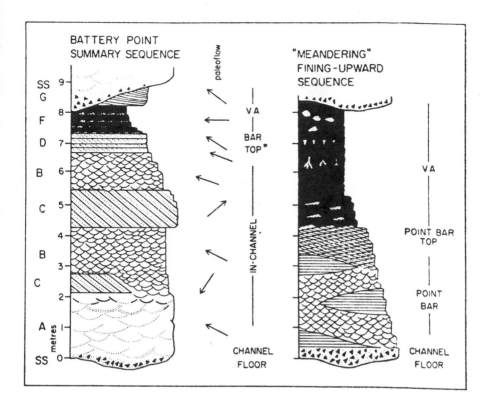

Figure 16   Summary sequence of Battery Point
Formation compared with meandering
river model (from Cant and Walker
1976, their Figure 16)

meandering. Rivers of this type presumably fall into the
"mixed-load" category, but the type of valley alluvium
indicated by Schumm (1968b, Figure 3) for these rivers
seems improbable. More likely is a deposit with characteristics intermediate between those of bedload (braided) and
suspended load (meandering) type.

    An ancient fluvial succession of this nature is ex-
posed in sections of the Devonian Peel Sound Formation at
Cape Anne, Somerset Island, NWT (Gibling and Rust, 1977).
Markov analysis shows a repetition of the sequence:
erosional scour with mudstone intraclasts, large-scale
trough cross-strata, small-scale trough cross-strata, and
plane-laminated sandstone, or siltstone/mudstone (Figure
17). On average, the latter lithology makes up 15% of the
succession; lateral continuity is good. Considered
separately, this section would probably be interpreted as
meandering, but the presence of numerous minor scours and
vertical transition into a section very similar to the
Malbaie Formation sandstone (Gibling, personal communication
1977) implies a distal braided environment.

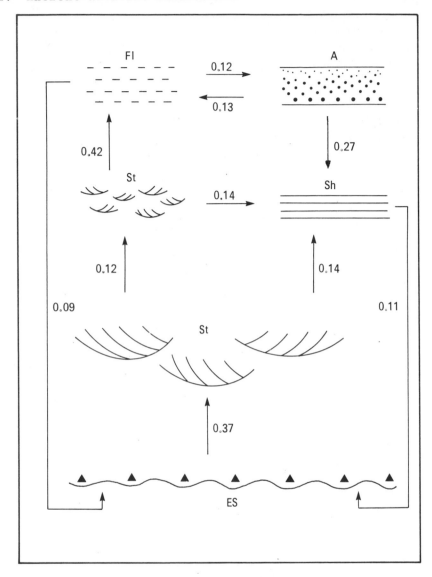

Figure 17   Summary diagram for the Peel Sound
Formation at Cape Anne, Somerset Island.
The numbers are transition probabilities
derived from Markov analysis of the
succession, which has 75 transitions
between "states". A 1.O probability between
two states indicates that they always occur
in sequence.
A: massive graded pebbly sandstone to
        sandstone
Sr: ripple cross-laminated sandstone;
        other states are as indicated in
        Figures 12 and 14

Paleohydrology of braided systems

The estimation of hydraulic and hydrological characteristics
of ancient braided systems is particularly difficult. The
channels are laterally unstable, and are therefore rarely
preserved intact in ancient deposits. If paleochannel
characteristics can be reconstructed, there is no way of
knowing whether it was a single major channel or one of
several. Large braided rivers confined within valleys
commonly have a single dominant channel, but this is not
always so (Rust 1972a, p 223). A single channel is unlikely to be dominant where the braided system spreads over
a wide flat area.

Baker (1974) attempted paleohydraulic estimates for
Quaternary alluvial terraces in Colorado. He assumed that
the terrace slopes approximate the Quaternary paleoslope,
and from the mean diameters of the largest particles was
able to derive depths of flow. Quaternary flood discharges
were then estimated to have been an order-of-magnitude
greater than modern flood discharges.

In most pre-Quaternary deposits, paleoslopes cannot be
determined directly because of the possibility of subsequent
tilting. Cant and Walker (1976) approached this problem by
estimating the Battery Point Formation paleoslope from the
Chezy equation, using a resistance coefficient appropriate
for dunes. The depth of the deepest scour was taken as an
indication of bankfull depth, and mean flow velocity at
this stage was determined from depth/velocity relationships
for dunes given by Harms and others (1975). From the
paleoslope the minimum bankfull discharge required for
braiding can be determined from the empirical relationship
established by Leopold and Wolman (1957).

Miall (1976) estimated sinuosities of paleochannels
in a Cretaceous braided stream deposit on Banks Island,
NWT from detailed measurements of planar cross-bed
orientation in vertical sections. The orientation of each
cross-bed set was assumed to define the orientation of its
respective bar; between-bar directional variance was then
obtained by summing the squared cross-bed deviations about
the moving average of the mean azimuth (Miall 1976, p 466).
Sinuosity could be determined by comparing variance with
that of modern rivers, or from an empirical formula
relating angular change of channel orientation to sinuosity
(Langbein and Leopold 1966). Estimates of paleohydrologic
parameters were based on determining water depth from the
magnitude of sedimentary structures (Miall 1976, p 477).
Using sinuosity and depth, width was obtained from an
equation relating sinuosity to channel width/depth ratio
(Schumm 1972, p 104), and mean flood velocities from
critical transport velocities for the largest clasts
present. In this way estimates of mean annual discharge
and mean flood discharge were obtained (Miall 1976, p 478).

Some of the problems associated with estimates of
this sort were outlined above. Another is that few of the
rivers on which Schumm's equations are based are braided.
It is difficult to determine the errors involved in each

Table 2:
Summary characteristics of braided alluvium

	Framework gravel	Sand
Proximal high slope	Imbricate, horizontal bedding >>planar cross-bedding Lithological variety: rapid fining downslope, interbedded debris flows Ex: Modern alluvial fans, L. Member, Cannes de Roche Formation	Abundant erosion surfaces with mudstone intraclasts; primary mudstone very rare. LAPL > trough cross-bedding. Great lateral variability; lacks cycles. Ex: Malbaie Formation sandstones
Proximal low slope	Imbricate, horizontal bedding >>planar cross-bedding. Relatively homogeneous lithology. Ex: Proximal Donjek outwash, Malbaie Formation conglomerates	Moderately developed cycles. Trough cross-bedding > planar cross-bedding. Minor mudstone. Ex: S.Saskatchewan River, Battery Point Fm.
Distal	Trough cross-bedding > horizontal bedding. Fining-upward cycles. Minor primary mudstone. Ex. Distal Donjek outwash, U. Member, Cannes de Roche Formation	Well developed cycles, good lateral continuity Ex: Peel Sound Fm Transitional to meandering deposits

calculation, and the errors could be cumulative. However, the estimates are a useful step towards understanding ancient alluvial systems, and can probably give reasonable indications of the size of the system involved. For many purposes this can be useful information, for example in the prediction and tracing of uranium or hydrocarbon reserves in alluvial deposits. Clearly, however, further knowledge of modern braided systems is required before our understanding of their ancient equivalents can be extended much further.

CONCLUSIONS

1. A simple classification proposed for alluvial channel patterns is based on two characteristics, determined at medium stage, and averaged for a given reach: channel sinuosity, and a parameter describing braiding intensity. This gives four categories, two of which are common: high sinuosity single-channel (meandering) and low sinuosity multi-channel (braided) systems.

2. The fining-upward model is useful for understanding meandering deposits and estimating dimensions and hydrologic characteristics of their paleochannels. There is need, however, for more data on variants, particularly coarser grained deposits.

3. Braided alluvial systems are more diverse, and their data base is less complete; existing sedimentary models are therefore inadequate for sophisticated analysis, such as paleohydraulics. There are three lithological types: silt-, sand-, and gravel-dominated. The latter two are the most abundant; their characteristics are summarised in Table 2, and they can be distinguished from meandering equivalents as follows: braided gravels are predominantly of framework type, while proximal sandy deposits preserve mud as intra-clasts rather than primary strata. Distal sandy braided deposits are transitional to, and may be indistinguishable from, deposits of meandering systems.

4. Longitudinal bars, the basic structure of gravelly braided systems, are regarded as primary bedforms, which exist in equilibrium with flood flows capable of transporting the total range of bed material.

ACKNOWLEDGEMENTS

I would like to thank M. R. Gibling, E.H. Koster and A. D. Miall for comments on the manuscript, H. J. Campeau for typing, and E. W. Hearn for drafting. Financial assistance from the National Research Council of Canada is gratefully acknowledged.

REFERENCES CITED

Alcock, F.J., 1935, Geology of Chaleur Bay region. *Geological Survey Canada Mem.*, 183, 146pp

Allen, J.R.L., 1963, The classification of cross-stratified units, with notes on their origin. *Sedimentology*, 2, 93-114

Allen, J.R.L., 1964, Studies in fluviatile sedimentation: six cyclothems from the Lower Old Red Sandstone, Anglo-Welsh Basin. *Sedimentology*, 3, 163-198

Allen, J.R.L., 1965, A review of the origin and character-istics of recent alluvial sediments. *Sedimentology*, 5, 89-191

Allen, J.R.L., 1970, A quantitative model of grain size and sedimentary structures in lateral deposits. *Geological J.*, 7, 129-146

Allen, J.R.L., 1973, Phase differences between bed configu-ration and flow in natural environments, and their geological relevance. *Sedimentology*, 20, 323-329

Baker, V.R., 1974, Paleohydraulic interpretation of Quaternary alluvium near Golden, Colorado. *Quaternary Research*, 4, 94-112

4. Ancient alluvial successions

Blissenbach, E., 1954, Geology of alluvial fans in semiarid regions. *Geological Society America Bull.*, 65, 175-190

Bluck, B.J., 1971, Sedimentation in the meandering River Endrick. *Scottish J. Geology*, 7, 93-138

Bluck, B.J., 1974, Structure and directional properties of some valley sandur deposits in southern Iceland. *Sedimentology*, 21, 533-554

Boothroyd, J.C. & Ashley, G.M., 1975, Processes, bar morphology and sedimentary structures on braided out-wash fans, northeastern Gulf of Alaska. In: Jopling, A.V. & McDonald, B.C., (eds), *Glaciofluvial and glacio-lacustrine sedimentation*. Soc. Econ. Paleontologists and Mineralogists Special Publication 23, 193-222

Bretz, J.H., Smith H.T.U. & Neff, G.E., 1956, Channeled scabland of Washington: new data and interpretations. *Geological Society America Bull.*, 67, 957-1049

Brice, J.C., 1964, Channel patterns and terraces of the Loup Rivers in Nebraska. *US Geol Survey Prof. Paper* 422-D, 1-41

Bull, W.B., 1972, Recognition of alluvial-fan deposits in the stratigraphic record. In: Rigby, J.K. & Hamblin, W.K. (eds), *Recognition of ancient sedimentary environments*. Soc. Econ. Paleontologists and Mineralogists Special Publication 16, 63-83

Cant, D.J., 1975, Sandy braided stream deposits in the South Saskatchewan River. *Geological Society America Abstracts with Programme*, 7, 731

Cant, D.J., 1977, A comparison of recent and ancient sandy braided river sedimentation. *Geological Association Canada, Programme with Abstracts*, 2, 10

Cant, D.J. & Walker, R.G., 1976, Development of a braided-fluvial facies model for the Devonian Battery Point Sandstone, Québec. *Canadian J. Earth Sciences*, 13, 102-119

Carlston, C.W., 1965, The relation of free meander geometry to stream discharge and its geomorphic implications. *American J. Science*, 263, 864-885

Chien Ning, 1961, The braided stream of the lower Yellow River. *Scientia Sinica (Peking)*, 10, 734-754

Church, M. & Gilbert, R., 1975, Proglacial fluvial and lacustrine environments. In: Jopling, A.V. & McDonald, B.C. (eds), *Glaciofluvial and glaciolacustine sedimentation*. Soc. Econ. Paleontologists and Mineralogists Special Publication 23, 22-100

Coleman, J.M., 1969, Brahmaputra River: channel processes and sedimentation. *Sedimentary Geology*, 3, 129-239

Collinson, J.D., 1970, Bedforms of the Tana River, Norway. *Geografiska Annaler*, 52A, 31-55

Conaghan, P.J. & Jones, J.G., 1975, The Hawkesbury Sandstone and the Brahmaputra: a depositional model for contin-ental sheet sandstones. *J. Geological Society Australia*, 22, 275-283

Costello, W.R. & Walker, R.G., 1972, Pleistocene sedi-
     mentology, Credit River, southern Ontario: a new
     component of the braided river model.   *J. Sedimentary
     Petrology,* 42, 389-400

Fahnestock, R.K., 1970, Morphology of the Slims River.
     Icefield Ranges Research Project, Scientific Results,
     1, 161-172

Fahnestock, R.K. & Bradley, W.C., 1973, Knik and Matanuska
     Rivers, Alaska: a contrast in braiding.   In:
     Morisawa, M. (ed), *Fluvial Geomorphology,* State Univ.
     of New York, Binghamton, 4, 220-250

Galay, V.J. & Neil, C.R., 1967, Discussion of "Nomenclature
     for bed forms in alluvial channels".   *J. Hydraulics
     Division ASCE,* 93, 130-133

Gibling, M.R. & Rust, B.R., 1977, Proximal and distal
     sandy braided alluvium in Devonian successions of the
     Arctic and Gaspé.   *Geological Association Canada,
     Programme with Abstracts,* 2, 20

Harms, J.C., Southard, J.B., Spearing, D.R. & Walker, R.G.,
     1975, Depositional environments as interpreted from
     primary sedimentary structures and stratification
     sequences.   Soc. Econ. Paleontologists and
     Mineralogists, Short Course Note 2, 161pp

Hein, F.J. & Walker, R.G., 1977, Bar evolution and develop-
     ment of stratification in the gravelly, braided,
     Kicking Horse River, British Columbia.   *Canadian J.
     Earth Sciences,*14, 562-570

Jackson, R.G., 1976, Depositional model of point bars in
     the lower Wabash River.   *J. Sedimentary Petrology,*
     46, 579-594

Johnson, A.M., 1970, *Physical processes in geology.*
     Freeman, San Francisco, 577pp

Langbein, W.B. & Leopold, L.B., 1966, River meanders -
     theory of minimum variance.   *US Geol. Survey Prof.
     Paper* 422-H

Leeder, M.R., 1973a, Sedimentology and paleogeography of
     the Upper Old Red Sandstone in the Scottish Border
     Basin.   *Scottish J. Geology,* 9, 117-144

Leeder, M.R., 1973b, Fluviatile fining-upwards cycles and
     the magnitude of palaeochannels.   *Geological Magazine,*
     110, 265-276

Leeder, M.R., 1975, Palaeogeographic significance of
     pedogenic carbonates in the topmost Upper Old Red
     Sandstone of the Scottish border basin.   *Geological J.,*
     11, 21-28

Leopold, L.B. & Wolman, M.G., 1957, River channel patterns:
     straight, meandering and braided.   *US Geol. Survey
     Prof. Paper* 282-B, 39-85

Leopold, L.B., Wolman, M.G. & Miller, J.P., 1964, *Fluvial
     processes in geomorphology,* Freeman, San Francisco,
     522pp

4. Ancient alluvial successions

Malde, H.E., 1968, The catastrophic Late Pleistocene Bonne-
ville Flood in the Snake River Plain, Idaho. *US
Geol. Survey Prof. Paper* 596

McDonald, B.C. & Banerjee, I., 1971, Sediments and bed forms
on a braided outwash plain. *Canadian J. Earth Sciences,*
8, 1282-1301

McGerrigle, H.W., 1950, The geology of Eastern Gaspé.
Quebec Department Mines, Geological Report 35, 168pp

McGowen, J.H. & Garner, L.E., 1970, Physiographic features
and stratification types of coarse-grained point bars:
modern and ancient examples. *Sedimentology,* 14, 77-112

McGregor, D.C., 1973, Lower and Middle Devonian spores of
Eastern Gaspé, Canada. 1. Systematics. *Palaeonto-
graphica,* 142, 1-77

McKee, E.D., Crosby, E.J. & Berryhill, H.L., 1967, Flood
deposits, Bijou Creek, Colorado, June 1965.
J. Sedimentary Petrology, 37, 829-851

Miall, A.D., 1976, Paleocurrent and paleohydrologic analysis
of some vertical profiles through a Cretaceous braided
stream deposit, Banks Island, Arctic Canada.
Sedimentology, 23, 459-483

Miall, A.D., 1977, A review of the braided-river depositional
environment. *Earth-Science Reviews,* 13, 1-62

Moody-Stuart, M., 1966, High- and low-sinuosity stream
deposits, with examples from the Devonian of Spits-
bergen. *J. Sedimentary Petrology,* 36, 1102-1117

Nami, M., 1976, An exhumed Jurassic meander belt from
Yorkshire, England. *Geological Magazine,* 113, 47-52

Ore, H.T., 1964, Some criteria for recognition of braided
stream deposits. *Univ. Wyoming Contribution Geology,*
3, 1-14

Padgett, G.V. & Ehrlich, R., 1976, Paleohydrologic analysis
of a late Carboniferous fluvial system, southern
Morocco. *Geological Society America Bull.,* 87,
1101-1104

Pettijohn, F.J., 1975, *Sedimentary rocks,* Harper and Row,
New York, 628pp

Reineck, H.-E. & Singh, I.B., 1975, *Depositional sedimentary
environments, with reference to terrigenous clastics.*
Springer-Verlag, New York, 439 pp

Rust, B.R., 1972a, Structure and process in a braided river.
Sedimentology, 18, 221-245

Rust, B.R., 1972b, Pebble orientation in fluvial sediments.
J. Sedimentary Petrology, 42, 384-388

Rust, B.R., 1975, Fabric and structure in glaciofluvial
gravels. In: Jopling, A.V. & McDonald, B.D. (eds),
Glaciofluvial and glaciolacustrine sedimentation.
Soc. Econ. Paleontologists and Mineralogists, Special
Publication 23, 238-248

Rust, B.R., 1976, Stratigraphic relationships of the
    Malbaie Formation (Devonian), Gaspé, Quebec.
    Canadian J. Earth Sciences, 13, 1556-1559

Rust, B.R., 1977a, The Malbaie Formation: sandy and
    conglomeratic proximal braided alluvium from the Middle
    Devonian of Gaspé, Quebec. *Geological Society America
    Abstracts with Programme,* 9, 313-314

Rust, B.R., 1977b, The Cannes de Roche Formation:
    Carboniferous alluvial fan and floodplain deposits in
    eastern Gaspé. *Geological Association Canada,
    Programme with Abstracts,* 2, 46

Schumm, S.A., 1963, Sinuosity of alluvial rivers on the
    Great Plains. *Geological Society America Bull.,*
    74, 1089-1099

Schumm, S.A., 1968a, River adjustment to altered hydrologic
    regimen-Murrumbridge River and paleochannels, Australia.
    US Geol. Survey Prof. Paper 598, 65 pp

Schumm, S.A., 1968b, Speculations concerning paleohydrologic
    controls of terrestrial sedimentation. *Geological
    Society America Bull.,* 79, 1573-1588

Schumm, S.A., 1972, Fluvial paleochannels. In: Rigby,
    J.K. & Hamblin, W.K. (eds), *Recognition of ancient
    sedimentary environments.* Soc. Econ, Paleontologists
    and Mineralogists, Special Publication 16, 98-107

Smith, D.G., 1973, Aggradation of the Alexandra-North
    Saskatchewan River, Banff Park, Alberta. In:
    Morisawa, M. (ed), *Fluvial geomorphology.* State
    Univ. New York, Binghamton, 4, 201-220

Smith, N.D., 1970, The braided stream depositional environ-
    ment: comparison of the Platte River with some
    Silurian clastic rocks, north-central Appalachians.
    Geological Society America Bull., 81, 2993-3014

Smith, N.D., 1971, Transverse bars and braiding in the
    lower Platte River, Nebraska. *Geological Society
    America Bull.,* 82, 3407-3420

Smith, N.D., 1974, Sedimentology and bar formation in the
    upper Kicking Horse River, a braided outwash stream.
    J. Geology, 82, 205-223

Walker, R.G., 1976, Facies models 3. Sandy fluvial systems.
    Geoscience Canada, 3, 101-109

Williams, P.F. & Rust, B.R., 1969, The sedimentology of a
    braided river. *J. Sedimentary Petrology,* 39, 649-679

Visher, G.S., 1972, Physical characteristics of fluvial
    deposits. In: Rigby, J.K. & Hamblin, W.K. (eds),
    Recognition of ancient sedimentary environments.
    Soc. Econ. Paleontologists and Mineralogists,
    Special Publication 16, 84-97

Part 2   SEDIMENT YIELD AND

WATER QUALITY

5 INTERDISCIPLINARY RESEARCH ON RUNOFF AND EROSION PROCESSES IN AN ARID AREA, SDE BOKER EXPERIMENTAL SITE, NORTHERN NEGEV, ISRAEL

*Aaron Yair

ABSTRACT

The paper describes a research program adopted by an inter-disciplinary team working on runoff and erosion processses, at an experimental watershed covering some 11 325 m^2. Instrumentation of the site consists of devices for measuring rainfall, wind, evaporation, soil moisture, runoff rate and yield, sediment (mineral and dissolved) rate and yield, amounts and distribution of biotic sediment, population dynamics of Isopods (woodlice) and the input of dust particles. The approach and instrumentation used for the study of each of the various aspects included in the research program are analysed in detail. Data obtained so far allow a separate study of each of the factors listed above as well as an analysis of the complex relationships existing among them. The layout and number of operating instruments (over 100) allow a close study of temporal and spatial variations in the processes and factors controlling them. Finally some benefits which emerge from the inter-disciplinary approach for geomorphic research are indicated.

INTRODUCTION

Slope runoff and erosion processes are presently being studied at an experimental site covering some 11 325 m^2, located in the arid Northern Negev (Figure 1). A strikingly non-uniform contribution of sediment from contiguous plots has been observed (Yair 1974). An analysis of rainfall, of runoff, and of surface properties such as slope length and gradient shows that these factors can hardly account for spatial non-uniformity in the delivery of sediment. Detailed field observations drew attention to intense digging and burrowing activity by desert animals such as porcupines and isopods (woodlice). Digging by porcupines seeking bulbs for their nourishment breaks up the soil crust which, due to its mechanical properties and biological cover of soil lichens, inhibits soil erosion. Thus fine soil particles with loose small aggregates are made available for transport by shallow flows. Similarly

*   Department of Geography, The Hebrew University, Jerusalem, Israel

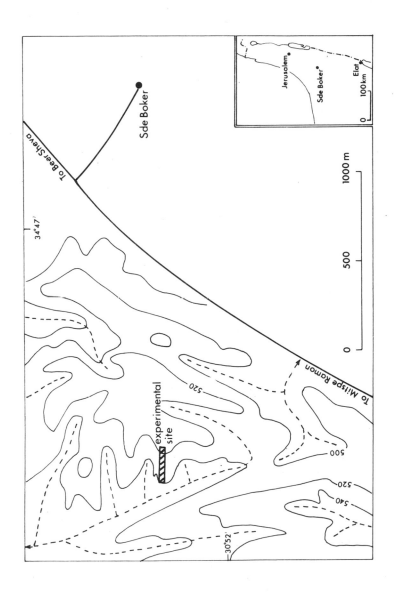

Figure 1 Location map

burrowing by isopods deliver small faeces
(Figure 2) which disintegrate easily under the impact of
raindrops. Following these observations measurements were
made of the amounts of available sediment produced through
biological activity. The data thereby obtained show that
these amounts are of the same order of magnitude as those
removed from the experimental site during a single rainy
season. Furthermore, the spatial variation in the amounts
of available sediment was found to conform to the observed
amounts of eroded material. Both patterns appear to be
very closely related to the microenvironments created by
the various lithological types of limestone outcropping
within the experimental site.

Following these results and complementary field
observations, an interdisciplinary research approach was
adopted in order to gain a better understanding of the
abiotic and biotic factors controlling runoff and erosion
processes in an arid limestone environment. The scientific
team working at the site presently includes a climatologist,
zoologist, botanist, pedologist and several geomorphologists.

The instrumentation of the site consists of devices
for measuring rainfall, wind, evaporation, soil moisture,
runoff rate and yield, sediment (mineral and dissolved)
rate and yield, amounts and distribution of biotic sediment,
population dynamics of isopods and the input of dust
particles. Data obtained so far allow a separate detailed
study of each of the factors listed above, and an analysis
of the complex relationships existing among them.

Aim of the present paper

This paper presents the various topics under current study,
describes the approach and instrumentation used and indicates the complex relationships which exist among the
various aspects studied. It is hoped that this paper will
clearly indicate the great theoretical and practical benefits afforded to geomorphological research by the adoption
of an interdisciplinary-ecosystematic approach.

SDE BODER EXPERIMENTAL SITE

Description of the site

The experimental site is located in the Northern Negev
(Figure 1). Average annual precipitation, based on a
record of 27 years, is 95 mm. However, annual rainfall may
vary from 34 mm to 167 mm. The number of rain-days varies
from 15 to 42, and only a few rains yield more than 20 mm
per day.

The site covers an area of 11 325 m^2. The drained
area is limited to one half of a first order drainage basin
extending over one side of the channel (Figure 3) facing
the North. The relative relief is about 30 m. Length of
slope varies from 55 m to 76 m and mean slope gradient,
as measured in the central part of plots, varies from 11.5%
to 29.5%.

The stratigraphic section is Turonian (Arkin & Braun
1965), represented here by the Drorim, Shivta and Netser

Figure 2   A mound of isopod pellets lying over
           a crusted topsoil

formations (Figure 2). Strata are subhorizontal with a
gentle dip of 4° towards the NNW. Although the three
formations are composed of limestone rock, they create
three different environments principally because of
difference in structure.

The Drorim formation outcrops in the lower part of the
experimental site. Rock strata are 10-30 cm thick and
very densely jointed. The rock weathers to cobbles and
boulders covering most of the surface (Figure 4). Bare
rock is exposed over about 20% of the outcrop, mostly at
the top of the formation. The soil cover is relatively
extensive, thick and stony. It is a desert brown lithosol
(Dan *et al* 1972) becoming a loessial serozem towards the
base of the slope. Repetitive wetting and drying cycles
produced a compacted topsoil crust. Pedologic differen-
tiation into defined soil horizons is very poor. Vegetation
is sparse. A cover of 5-10% is quite uniformly spread over
the entire area. The characteristic plant association is
that of *Artemisia herba alba - Gymnocarpos decander*
(Danin 1970).

The Shivta formation, a massive limestone, is exposed
over the central part of the site (Figure 3). Strata
thickness is 30 cm to 80 cm, forming a stepped topography

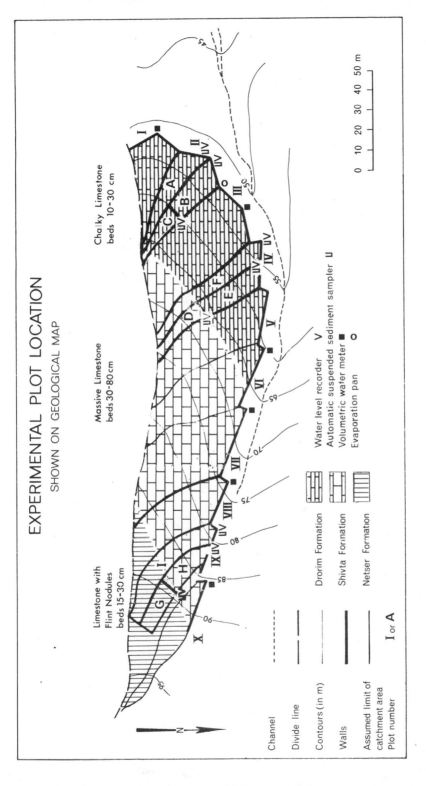

EXPERIMENTAL PLOT LOCATION

SHOWN ON GEOLOGICAL MAP

Limestone with Flint Nodules beds 15-30 cm

Massive Limestone beds 30-80 cm

Chalky Limestone beds 10-30 cm

Channel

Divide line

Contours (in m)

Walls

Assumed limit of catchment area

Plot number   I or A

Drorim Formation

Shivta Formation

Netser Formation

Water level recorder   V

Automatic suspended sediment sampler   U

Volumetric water meter ■

Evaporation pan O

0   10   20   30   40   50 m

Figure 3 Layout of runoff and erosion plots

113

Figure 4   Drorim formation - surface characteristics

Figure 5   Shivta formation - surface characteristics

with rock terraces (Figure 5). Bedrock is exposed over 50% of the surface. Jointing is spaced. The soil cover is represented by non-contiguous soil strips at the bases of rock terraces. The soil, a loessial serozem, is shallow, not exceeding 30 cm in thickness. A relatively rich cover of lichens and mosses, especially well-developed at the base of rock terraces, is characteristic of the Shivta formation. Thicker "soil" up to 80 cm in depth may be found in soil columns, filling the joints in the bedrock. This soil may be easily distinguished from that of the soil strips by its characteristic red colour, looseness and richness in organic matter. The vegetation cover, 5-10%, is concentrated along the soil strips and soil-filled joints. The characteristic plant association is that of *Varthemia montana - Organum dayi* (Danin 1970) which is indicative of a better soil moisture regime than that found over the Drorim formation. It is worthwhile noting that the above plant association is widely distributed over the Galilee Mountains where annual rainfall varies between 500 mm to 1000 mm.

The Netser formation is exposed at the upper part of the site. It is composed of limestone with flint concretions. Strata thickness is 10-30 cm with very dense jointing (Figure 6). Shattered rock is exposed over about 30% of the surface, yet the soil is very shallow and appears in numerous soil patches. The characteristic plant association is that of *Hammada scoparia,* indicative of the poorest soil moisture regime within the studied area.

General approach

Due to the strong influence of the hydrological approach in the study of watershed hydrology, most studies conducted by geomorphologists in instrumented watersheds were focused on channel processes. The systematic study of spatial variations in runoff and sediment delivered by the slopes was often ignored or given little attention. Such an approach may be acceptable in humid areas where overland flow is assumed to contribute negligible amounts to storm runoff. However, it would not be acceptable in arid areas where overland flow is more frequent than channel flow (Yair 1972; Yair & Klein 1974) and represents an essential stage in the initiation and development of floods. For this reason attention was focused on slope rather than channel processes at the Sde Boker experimental watershed.

A non-uniform slope contribution to runoff and sediment is to be expected within a watershed. Any watershed, even a lithologically homogeneous one, is characterised by systematic spatial variations in the topographic slope properties. Slopes are usually gentle and short near the upper divide. At the central part of the watershed they tend to be long and steep, while at the lower part again short and gentle. These topographic differences cause spatial variations in other surface properties such as depth, moisture regime, grain size and mineralogic composition of the soil, which affect abiotic as well as biotic factors involved in runoff and erosion processes. Such differences, combined with the non-uniform rainfall distribution observed in small watersheds (Cappus 1965; Sharon 1970; Shanon 1976) cause significant spatial differences in slope runoff and sediment delivery which certainly merit a systematic study.

ANALYSIS OF TOPICS STUDIED

Rainfall characteristics*

General approach

Rainfall is an essential factor in the ecosystem studied, as it directly controls runoff and erosion processes and indirectly controls - through soil moisture - the spatial distribution of the activity of digging and burrowing animals, as well as the distribution of plant associations. Consequently special attention has been given to the obtaining of accurate data concerning spatial variation of the effective rainfall intercepted by the surface of slopes with varying aspects and inclination. In studying effective rainfall reaching the ground, two different aspects should be considered: that of the meteorologist and that of the researcher concerned with rain-conditioned processes taking place on the ground surface. Although closely related, these two aspects have rarely been investigated simultan-eously. The first approach, that of the meteorologist, is based upon the use of conventional raingages with hori-zontal orifices. Such raingages measure the amount of rainfall passing through the lowest part of the atmosphere above a given point on the ground. Spatial differences in rainfall amounts collected reflect differences in yield of clouds or in the density of the falling rain before reaching the ground. The amount thus collected is entirely independent of the geometric position - ie. aspect and inclination - of the ground surface on which the raingage is located. The second approach, that of the geomorphologist studying dynamic processes, or the ecologist, is mainly concerned with the actual amount of rainfall intercepted on the ground surface. As rain mostly falls obliquely, this amount - like solar radiation - depends upon the local slope angle and the aspect of the receiving ground surface in relation to the paths of falling raindrops. In middle latitudes, inclinations up to 40° measured from the zenith have been reported by Leyert (1959) at a flat area in Holland, and values above 50° are common in mountainous areas (Hamilton 1954; Grunow 1953). In order to measure the rainfall amount actually intercepted on the ground surface, the above mentioned investigators used raingages with their orifices lying in a plane exactly parallel to the sloping surface concerned. Thus, the orifices constituted a truly representative sample of the ground surface, with respect to local aspect and slope inclination. While slope angle and aspect are constant during the time that data is accumulated, the inclination of falling raindrops may vary from storm to storm depending upon local wind conditions,

*Conducted by D. Sharon

Figure 6 Netser formation - surface characteristics

Figure 7 General view of plot II

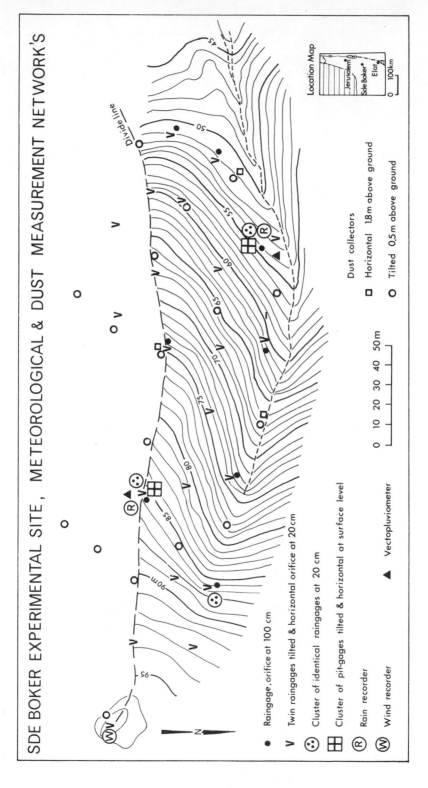

Figure 8 Layout of meteorological instrumentation and dust collectors

which therefore play an important role in non-uniform
rainfall distribution.

Taking into account the above considerations, spatial
variations of geomorphologically effective rainfall are
much larger than had been assumed previously. Considering
prevalent values of slope gradient (20°) and rainfall
inclination (40°) within the experimental site, a windward
facing slope receives almost twice as much rainfall as an
adjacent leeward-facing one. For slopes of 15°, the ratio
is about 1:1.5. It is important to note that as storm
duration is uniform over small areas, the above differences
in rainfall amounts actually reflect differences in rainfall intensities, thus increasing the effect of non-uniform
rainfall distribution on runoff and erosion processes.

Experimental design and procedure

In order to construct a mathematical model for prediction
of effective rainfall, combining all factors discussed
above, the following experimental design (Figure 8) and
procedure were adopted:

A network of 21 stations has been established, each
station includes twin raingages: one tilted (orifice
parallel to the ground) and the other with an
horizontal orifice, each with an orifice area of
about 6.5 cm^2, installed 30 cm above the ground. In
order to obtain the spatial variation of rainfall,
the raingages were distributed along three lines:
along the base of the slope area, at mid slope, and
along the local divide line. In addition, two control
stations were installed one on the ridge and the
other near the channel. Each of these stations
includes the following equipment: a pit gage, 200 cm^2
in area, with a horizontal orifice at ground level
after Rodda (1967) and Green (1970) for estimating
wind effect loss in raised raingages. Three identical
horizontal raingages at 30 cm above the ground and
about 100 cm apart, which estimate the sampling
error of point measurement for each storm. A Hellman
type rainfall recorder.

Wind speed and direction are measured at 3.5 m above
the ground by a Woelfle recorder at the top of the
hill. Thus, simultaneous records of wind and rainfall
are available. In addition, several micro-meteorolo-
gical experiments were conducted at 10 stations for
studying local topography-conditioned wind flow
patterns.

Raingage observations are made shortly after each storm.
During the initial phase of the study, observations were
recorded from all above mentioned equipment. Based on data
collected, the trigonometrical relationship between the
catch in standard (horizontal) and inclined raingages was
studied, considering wind speed and direction. Once this
relationship is established, spatial distribution of
incident rainfall will be easily estimated from a much
smaller amount of data, allowing for a substantial reduction
in the number of operating raingages.

5. The Sde Boker Experimental Site

Rainfall-runoff relationships*

General approach

Recent studies on the process of runoff generation in humid areas (Betson 1964; Hewlett & Hibert 1967; Amerman & McGuiness 1967; Kirkby & Chorley 1967; Betson & Marius 1969; Dunne & Black 1970; Freeze 1974; Moor & Taylor 1975; etc) have respectively shown, albeit in different ways, that the response of a small watershed to storm runoff is not uniform throughout its area, even if rainfall conditions are considered to be uniform. Such non-uniformity is attributed by all authors quoted to spatial variation in soil moisture conditions. The most effective areas contributing to storm runoff were the channel itself and a belt of varying width extending on both sides of the channel, in which soil was saturated or nearly saturated. Such a situation may result from the proximity of the ground water table to the valley bed; from the slow but steady movement of soil water towards the base of the slopes at a shallow depth below the sloping surface or from a combination of these two factors. The results obtained in the above quoted works led to the formulation of a general concept entitled "partial area contribution to storm runoff". A similar spatial variation in soil moisture would be unlikely in hot deserts where rainfall is scarce, evaporation is very high and the soil cover, where it exists, is shallow and patchy. Nevertheless, the basic assumption of the present study is that runoff generation, within the experimental site, follows a similar pattern of partial area contribution to that identified for humid areas. It is caused by factors which were overlooked in humid areas such as rainfall regime, surface properties, other than soil moisture, or a combination of these two sets of factors.

I   Cappus (1958), Sharon (1970) and Shanan (1976) showed that in virtually every individual storm, rainfall distribution over small watersheds is non-uniform. A larger amount of rain has been found to fall in the channel and adjoining slopes than over the ridges. Such a rainfall distribution pattern may cause a relatively quick generation of runoff at the base of the slopes. The process is probably enhanced by the fact that differences in rainfall amounts actually represent differences in intensity. Under such conditions, the frequency and magnitude of runoff events are expected to be higher at the bases of the slopes than at their upper parts.

II   Rainfall duration in the desert is often very short. Even a daily rainfall of 15 mm, which happens in the Sde Boker area about once a year, may be subdivided into four to six individual short showers separated by sunny or windy periods. When such showers, each totalling a few millimeters of rainfall, do generate runoff, water layer depth is very shallow. Considering the high roughness of the stony environment, flow velocities of shallow flows are extremely

---

* Conducted by A. Yair and H. Lavee

low, of the order of 3 cm to 6 cm/sec, very seldom exceeding 10 cm/sec (Emmett 1970; Lavee 1973). Under such conditions surface runoff if generated at the upper part of a long slope, probably has little to no chance of reaching the slope base. Therefore it could not contribute to storm channel runoff. On the other hand, runoff water which infiltrates into the slope itself is very important from the point of view of the ecologist. The infiltrated water improves the soil moisture regime of the upper and middle part of the slope creating microenvironments favorable to biological activity.

III Finally, the phenomenon and pattern of partial area contribution to storm runoff may also be controlled by surface properties linked with the local litho-stratigraphic section. Passage from a lithological unit yielding a high runoff rate at the upper part of the slope to a second unit located downslope and characterised by high permeability, may prevent run-off generated upslope from reaching the channel, considering the local nature of rainfall conditions. Such a situation often prevails in the study area. In the central part of the drained area, runoff generated upslope within the Shivta formation infiltrates down-slope in the outcrops of the Drorim formation before reaching the channel (Yair 1974).

Experimental design and procedure

In order to obtain data on spatial variations in slope runoff, the experimental site was initially subvidided into ten contiguous plots draining the entire slope area from the watershed divide to its mouth (Figure 3). Low walls constructed at the base of the slopes directed runoff to collectors which were emptied through volumetric water meters. Later on, plots II, IV and IX were each subdivided into three subplots: one long, extending from the divide to the slope base and the two adjoining ones short, draining the upper or lower sections of the slope (Figure 3; Figure 7). While plot II extends over a uniform lithological unit, plots IV and IX each cover two different units (Figure 3). Each of the subplots is equipped with a very sensitive water level recorder. In addition, soil moisture of the upper soil layer is measured at nine stations distributed throughout the site. Three soil samples are taken at each station immediately after a rain event and at subsequent intervals of two weeks. Evaporation is measured by a single recording pan located near the channel (Figure 3).

Available data on the distribution of effective incident rainfall, evaporation, soil moisture, runoff rate, runoff yield and surface properties will allow a detailed study of the factors controlling runoff generation over each of the plots. By comparing hydrographs obtained simultaneously at each of the subplots equipped with a stage recorder, it will be possible to analyse the degree of continuity of the runoff process along the slopes, as

well as the width of slope belt contributing runoff to
the channel under different rainfall conditions.   Such
analysis is possible for both uniform and non-uniform
lithological conditions.

     To complete the information obtained during natural
rainfall events, simulated rainfall experiments are planned.
The sprinkled area will cover plots II, IV and IX
individually.   Using tracing methods, flow velocities, flow
continuity and width of runoff contributing belt will be
measured.

Denudation processes

The study of denudation processes is presumably the most
complex due to the diversity of active processes and the
difficulty in controlling directly, under field conditions,
the factors involved in the mechanism of detachment of
soil particles at numerous points over the experimental
site.   The present study will be limited to the aspects of
erosion by surface runoff.   Creep and other mass movement
processes will be excluded from the study even though their
role in the modelling of the local landscape might be
quite important.   Sediment delivery rate and yield will be
related to runoff regime on the one hand, and to the
spatial distribution of loose available sediment prepared
by the foraging activity of animals on the other.   The
study of soil erosion by surface runoff will be supplemented
by that of chemical denudation.   An attempt will be made
to trace the movement of sediment along the slopes under
different rainfall conditions.   Finally, regarding rainfall
and runoff, the study will emphasize the aspect of
spatial variability in the sediment yield under different
rainfall conditions.

Soil erosion by surface runoff*

Collection of sediment data is based on the network of
16 runoff plots.   Those plots equipped with a stage
recorder are also equipped with an automatic suspended
sediment sampler (Figure 3) each containing 24 500 cc
bottles.   The first water sample is taken at the very
beginning of the flow and the following samples until the
flow ends at fixed intervals of 2.5 or 5 minutes.   Filling
the bottles takes ten seconds.   Samples collected are sub-
mitted to a standard concentration analysis. Figures
obtained, together with recorded runoff rate, allow an
accurate analysis of variation in sediment concentration
during each flow event, as well as computation of the
sediment yield for each of the nine plots.   In addition,
sediment data is being obtained for the plots equipped with
volumetric water meters (Yair 1974).

Chemical denudation by surface runoff**

Several factors led to the assumption that the rate of
chemical denudation may be of importance in the study area,
despite its aridity: a) $CaCo_3$, which represents the major
component   in the chemical denudation process is abundant

---

*Conducted by A. Yair and H. Lavee
**Conducted by A. Yair and R. Gerson

in the research area. All rock units are made of limestone
and the soil itself is rich in $CaCo_3$, accounting for 30-50%
of the soil's weight. b) Preliminary experiments conducted
at the experimental site show clearly that the highest concentrations in $CaCo_3$ are obtained at the very beginning of
the flow, decreasing rapidly within a few minutes to very
low values. The geomorphic significance of such results
is that even shallow and short flows, quite frequent in
the experimental site, are effective in terms of chemical
denudation. c) The loessial soil existing in the area is
mainly airborne as indicated by the results of numerous
soil mechanical analyses. These showed that 80-95% of the
soil is composed of silt and sand particles typical of
material transported and deposited by wind (Ginsburg &
Yaalon 1963). Under such conditions, removal of alloch-
tonous soil by surface runoff may only partly account for
a change in landscape on a geological scale. Such changes
presume the process of slope recession following weathering
and disintegration of the local bedrock. One may therefore
assume that, despite the arid prevailing climate, solution
of the limestone rock and detachment of blocks from bedrock
outcrops play an important role in the effective process
of slope recession and landscape evolution.

The study of chemical denudation began two years ago,
during which more than 500 water samples were collected.
Following each runoff event a volume of 250 cc is
extracted from each of the bottles of the suspended sediment samplers. The following ions are analysed quanti-
tatively: Ca^{++}, Mg^{++}, Na^{+}, K^{+}, Cl^{-} and HCO_3^{-}. On several
occasions, rainwater was analysed as well. Data obtained
show that most of the Ca^{++} and about half of the Mg^{+} are
derived from the complex rock-soil, whereas most of the
chlorides are supplied by rainwater.

The procedure adopted enables the analysis of the
following main aspects: 1) relative importance of chemical
denudation as compared to soil erosion under different
rainfall conditions over each of the microcatchments
equipped with stage recorders and sediment samplers;
2) variation during flow of concentrations in ions
analysed; 3) comparison of chemical denudation rate in the
study area to other carbonate terrains representing
different climatic environments.

Sediment yield and biological activity*

The important role assigned to sediment made available by
the foraging activity of animals is based on the combination
of three factors: a) isopod pellets and soil aggregates
provided by porcupines are obviously more easily eroded by
shallow flows than is crusted soil; b) the amounts of
"biotic sediment" are on the same order of magnitude as
the sediment amounts actually collected at the base of slope
plots; c) spatial distribution of the "biotic sediment"
is not uniform and is similar to that of the eroded
material.

There is a positive relationship between the extent
*Conducted by A. Yair, M. Shachak and H. Lavee

of foraging activity and the size of the population involved. Under the restrictive arid climatic regime, the size of such population is not constant and may vary greatly from one year to the next. For example, following a drought year, the number of active isopod burrows may drop by 50-90%, thus drastically limiting the amounts of available sediment prepared during the following hot season. Similarly, overexploitation of geophytic bulbs by porcupines during a given year will lead to a decrease in their activity during the following years until the regeneration of such bulbs. On the other hand, rapid changes in population size - mainly of isopods - may be related to purely biological-genetic factors. It thus appears that rainfall and runoff characteristics of a given season, together with biological factors, could have more influence upon the amounts of sediment removed from the slopes during the next season than could other factors such as slope length and gradient. This leads directly to the hypothesis that temporal variations in slope sediment delivery bear a cyclic form at least partially related to cycles of production of available sediment. Needless to say, a cyclic approach to the process of soil erosion differs radically from the current approach to the study of soil erosion.

Without denying the effect of available biotic sediment on sediment yield, its assumed predominant role may be questioned on the following grounds: 1) the similarity between the non-uniform sediment amounts collected at the base of slope plots and the distribution and amounts of loose available sediment may be a coincidence. It cannot be considered as a direct proof concerning the origin of the transported material. In fact, isopod pellets, as well as soil aggregates prepared by porcupines, disintegrate rapidly under the impact of raindrops and during the flow, thus preventing any possibility of differentiating between biotic and abiotic sediment. 2) The term "available" indicates a relative situation rather than an absolute or objective one. There can be no doubt that under shallow and low energy flows loose pellets and disaggregated soil are more easily entrained than the crusted soil (Figure 2). However, during runoff events of high magnitude the crusted soil, which covers much larger areas than those occupied by biotic sediment, becomes available to erosion. It may potentially contribute huge amounts of sediment greatly in excess of that prepared by animals. As the eroded soil is probably often of mixed biotic and abiotic origin, a suitable analysis of the role of the biological factor as a sediment contributor should therefore be conducted within the broader framework of the basic geomorphic problem of frequency and magnitude of erosion processes and their relative role in landscape modelling. 3) The availability of the loose sediment decreases with time during the winter season, due to the combination of two processes: a) part of the biotic sediment is washed away by the first flow events leaving smaller and smaller amounts on the surface; and b) the loose material left over the surface develops a thin soil crust because of wetting and drying cycles. This reduces the availability of the underlying

pellets or soil aggregates. It therefore appears that the
role of the biotic sediment is greater at the very
beginning of the rainfall season than afterwards when it
may be either gradually or sharply reduced to a negligible
one. 4) Finally, if the assumption concerning the partial
area contribution to storm runoff is correct for the study
area, the sediment yield at a given runoff event should be
compared with the amounts of available sediment lying
within the belt contributing water to the slope basis. At
the same time available sediment detached at the upper part
of the slope is deposited at its middle or lower part
following infiltration of runoff water. Such a process does
not immediately contribute available sediment to the channel
but participates in the overall process of sediment movement
along the slopes.

The following procedure was adopted for the quantitative mapping and monitoring of the movement of available
sediment. The experimental site was subdivided by a grid
into 204 squares with the long axis of the grid lines
parallel to the strike line of the geological formations.
A detailed survey is conducted each year at the end of the
summer. At each square porcupine digging points and isopod
burrows are counted and multiplied by the average weight of
sediment produced per burrow or digging point. Following
each runoff event a survey of the available sediment left
over the surface is conducted at plots II, IV and IX.
Sediment amounts collected at the plots are compared to the
amounts of available sediment which have been washed away.
Finally attempts are being made to monitor travel distances
of available sediment by means of suitable tracers.

Population dynamics of isopods*

General approach

A complete analysis of the biological factor from the
geomorphic-ecosystematic point of view should include
several aspects: as stated above, changes in population
size may affect the production of "biotic sediment". On
the other hand, the burrowing and digging activity of
animals contributes to the process of soil turnover; and
isopods, by adding organic matter to ingested organic
soil, contribute to the decomposition of organic matter
thus playing some role in the cycling of materials in the
desert environment. For the time being most of our
attention is focused on the first aspect, especially on the
population dynamics of isopods. This restriction was
imposed for two reasons: a) isopods represent a permanent
resident population while porcupines are only occasional
visitors; b) more scientific information is available on
the life history and burrowing strategy of desert isopods
(Shachak et al, 1976).

Studies on population dynamics are usually conducted
by zoologists or ecologists. While literature on the
influence of biological factors, such as predation and
competition, is relatively abundant, information concerning

---

*Conducted by A. Yair and M. Shachak

the influence of physical factors, at least in desert areas,
is rather incomplete. Especially no attempt has been made
to study spatial and temporal differences in the population
size of isopods in relation to spatial differences in lithological units. In order to elucidate the approach adopted
in the present study, which emphasizes the role of the
lithological factor, some basic information concerning isopod life is provided. Isopods are annual creatures living
in large families of 40 to 70 members each. In late winter
nine-month old isopods vacate their burrows, the females
begin to excavate new burrows, pair formation occurs,
families grow rapidly and digging activity increases
concomitantly. The new settlers may be subdivided into two
categories! a) Successful families - those which provide
most of the sediment - are capable of surviving through the
hot season. Their burrows are confined to those microenvironments in which soil depth exceeds 30 cm and soil
moisture content remains higher than 6% by weight at the
end of the long and hot summer. b) Unsuccessful families
whose activity stops in spring or summer due to deficient
soil moisture regime. It therefore appears that soil
moisture regime may be considered as a key physical factor
in the settlement and survival of isopod families. Some
positive relationship should therefore be expected between
the spatial distribution and density of isopod families and
that of soil moisture. Assuming a rather uniform evaporation rate, the soil moisture regime will depend upon two
sets of factors: the first, related to water input into
the soil; the second referring to the properties of the
soil cover as a medium capable of absorbing the input of
water and retaining it throughout the hot season. This
latter set of factors, as well as the threshold amount of
rainfall needed to generate runoff, are closely related to
ecological microenvironments which, in arid areas, differ
greatly from one rock unit to another.

Data obtained so far fully justify the approach
adopted. Surveys conducted during five consecutive
seasons show clearly that the density of isopod burrows is
always higher within the Shivta outcrops where moisture
conditions are best than over the Drorim and Netser
formation (Yair 1974). Furthermore, temporal fluctuations
in population size are apparently more limited in the
Shivta formation.

Experimental procedure

The grid used for estimation of available sediment also
serves for the study of isopod population. At each square,
successful and unsuccessful families are counted. Five
such surveys have been conducted since 1973. The data
thus collected allow an analysis of temporal fluctuations
in population size within each lithological unit and a
comparative study of population dynamics between lithological units for each of the survey years. Population
dynamics of isopods will be related to available data on
rainfall and runoff regime, as well as to soil properties,
including soil moisture regime. All this in an attempt to
define the specific role played by the physical factors of

the environment in observed changes of isopod population size.

Collection of dustfall*

General approach

The study area is located within the northern fringe of an arid climatic belt where deflation and wind deposition processes occur. The net balance between the amounts deposited by wind and those eroded by runoff and deflation processes is crucial to the understanding of present geomorphic trends. In a recent study devoted to the collection of dustfall throughout Israel, dustfall rates on the order of 250-300 gr/m²/yr were obtained by Ganor (1975) for the Northern Negev. Such figures are two or three times higher than known values on erosion rates by surface runoff over the same area. Unfortunately there is no data on deflation rates over the Northern Negev. If deflation rate is rather low because of the very existence of a mechanical and biological soil crust, it may well be that the study area is presently undergoing aggradation processes. In any case, considering the absolute amounts of dustfall involved, the input of airborne particles into an area where detailed studies on runoff and erosion processes are conducted merits to be studied quantitatively and its geomorphic contribution to present day processes defined.

The systematic study of dustfall over the experimental site began in April 1976. As with all other topics, special attention is accorded to spatial variations in dustfall rates. The lack of a suitable and reliable method for monitoring deflation processes has so far prevented the study of this complementary aspect whose importance remains unknown.

Experimental design and procedure

The measurement apparatus consists of 20 buckets filled with water for dust collection. Three such buckets installed at an elevation of 1.8 m above the ground have horizontal orifices and are intended to represent dust concentration in the air. In order to represent actual amounts of dust reaching the ground, 17 buckets with orifices parallel to the local ground surface were installed at 50 cm above the ground. The location of the buckets forms a grid (Figure 8) which allows the simultaneous study of spatial variations in dustfall over the slopes. Either once a month or following a dust storm, the buckets are carefully washed, the dust extracted and weighed after an oven drying process.

CONCLUSIONS

The objective of the present paper was not to present and discuss results already obtained but rather to describe the methodological approach adopted and indicate some of its advantages for geomorphic research. The prominent characteristics of the approach adopted may be summarized

*Conducted by A. Yair, D. Sharon, H. Lavee, D. H. Yaalon and E. Ganor

as follows:

> Various aspects pertaining to the problem of runoff
> and erosion processes are studied simultaneously at
> the same experimental site by an interdisciplinary
> team.
>
> Special emphasis is placed on obtaining a better
> understanding of spatial and temporal variations in
> the processes and factors controlling them. For
> this purpose very dense networks of appropriate
> instruments were installed. The total number of
> instruments installed over an area of 11 325 m^2 is
> close to one hundred; the number of laboratory
> analysis (sediment concentration, chemical, dust,
> etc.) performed exceeds one thousand per year. As
> the study of each topic is based on a working hypo-
> thesis, the spatial location of the instrumentation
> is never random.
>
> Although the approach adopted is interdisciplinary
> an attempt is made to isolate several key factors
> which are studied in a highly detailed manner. It is
> our belief that a good understanding of such factors,
> together with accurate data, precedes any valid
> attempt at understanding the complex relationships
> between two or more factors which often are not
> independent.
>
> Due to the interdisciplinary approach the biological
> factor, often overlooked by geomorphologists, was
> integrated into the study of erosion processes.

Although it may be premature to evaluate the entire
project, especially the methodological approach adopted,
one important benefit has emerged at the present research
stage. Discussions among members of the interdisciplinary
team prior to the study of a given factor led to the
adoption of methods and procedures specially designed for
interdisciplinary research which differ from those used
by each discipline individually. For example, rainfall is
not measured in conventional meteorological manner. Rather,
rain is measured in a manner more appropriate to the needs
of the geomorphologist or the ecologist who are interested
in the problems of rainfall-runoff relationships and soil
moisture regime. The pedologist who focuses his attention
on the soil cover spread over the bedrock is not really
interested in the properties of the soil filling the bedrock joints as well as the bedding planes between rock
strata. But this soil, which offers the most favorable
microenvironment for isopod burrows and geophytic bulbs
must be considered by an ecologist. Similarly, the study
of population dynamics of isopod families is no more based
on a random network but directly on the spatial network of
microenvironments related to the specific characteristics
of different lithological units. Finally, the introduction
of the biological factor may lead to a new approach in the
study of slope erosion processes, namely the cyclic one.
Obviously such an approach should be limited to those areas

where the delivery of available sediment is quantitatively significant.

In conclusion, geomorphology is usually considered as an interdisciplinary scientific discipline. However, it has not yet fully developed its proper tools and methodology. For the time being most geomorphologists are often satisfied with methods used by neighboring disciplines, borrowing them for their own research. It is our hope that the present study clearly shows the necessity of adapting some of these methods to the specific needs of geomorphic research. By doing so, new tools and new approaches may appear which, in the long run, may be very helpful for assessing the specific scientific contribution of our discipline in comparison with allied fields.

MEMBERS OF THE RESEARCH TEAM

Dr. A. Yair	Geomorphologist	Head of Project
Prof. D. Sharon	Climatologist	The Hebrew University
Mr. H. Lavee	Geomorphologist	The Hebrew University
Dr. M. Shachak	Zoologist	Desert Research Institute, Ben Gurion University of the Negev
Dr. R. Gerson	Geomorphologist	The Hebrew University
Dr. A. Danin	Botanist	The Hebrew University
Prof. D.H.Yaalon	Pedologist	The Hebrew University
Dr. E. Ganor	Climatologist	Tel Aviv University

ACKNOWLEDGEMENTS

Financial assistance of the Central Research Fund of the Hebrew University and of the Desert Research Institute, Ben Gurion University of the Negev, Sde Boker Campus, is kindly acknowledged. Thanks are accorded to Mr. H. Levy and Mr. Y. Trostler for their technical assistance, and to Mrs. T. Sofer for drawing the illustrations. It is a pleasure to thank Mr. E. Milo of the Desert Research Institute for his regular help in fieldwork and data collection.

REFERENCES CITED

Amermam, C.R. & McGuiness, J.L., 1967, Plot and small watershed runoff: its relation to larger areas. *Trans. American Society Agricultural Engineers,* 464-466

Arkin, Y. & Braun, M., 1965, Type sections of upper Cretaceous formations in the Northern Negev (Southern Israel). *Geological Survey - Stratigraphic Section 2a, Jerusalem.*

Betson, R.P., 1964, What is watershed runoff? *J. Geophysical Research,* 69(8), 1541-1552

Betson, R.P. & Marius, J.B., 1969, Source areas of storm runoff. *Water Resources Research,* 5(3), 574-582

5. The Sde Boker Experimental Site

Cappus, P., 1958, Répartition des précipitations sur un bassin versant de faible superficie. *Proc. General Assembly of Toronto 1957. International Association Scientific Hydrology,* Publication 43, 515-528

Dan, J., Yaalon, D.H., Koyumdhisky. H. & Z. Raz, 1972, The soil association map of Israel. *Israel J. Earth Sciences,* 21, 29-49

Danin, A., 1970, *A phytosociological-ecological study of the Northern Negev of Israel.* Unpublished PhD thesis. The Hebrew University. Mimeog. 214pp (Hebrew, English summary)

Dunne, T. & Black, R.D., 1970, Partial area contribution to storm runoff in a small New England watershed. *Water Resources Research,* 6(5), 1296-1311

Dunne, T., Moore, T.R. & Taylor, C.H., 1975, Recognition and prediction of runoff producing zones in humid regions. *Hydrol. Science Bull. XX,* 3(9), 305-327

Emmett, W.W., 1970, The hydraulics of overland flow on hillslopes. *US Geol. Survey Prof. Paper* 662-A

Freeze, R.A., 1974, Streamflow generation. *Reviews of Geophysics and Space Physics,* 12(4), 627-647

Ganor, E., 1975, Atmospheric dust in Israel. In: *Sediment-ological and meteorological analysis of dust deposition.* Unpublished PhD thesis. The Hebrew University, 224 pp

Ginzbourg, D. & Yaalon, D.H., 1963, Petrology and origin of the loess in the Beer-Sheva basin. *Israel J. Earth Sciences,* 12, 68-70

Green, M.J., 1970, Effects of exposure on the catch of rain gauges. *J. Hydrology,* 9(2), 55-71

Grunov, J., 1953, Niederschlagsmessungen am Hang. *Meteor. Rundschau,* 6(5/6), 85-91

Hamilton, E.L., 1954, Rainfall sampling on rugged terrain. *US Department Agriculture Technical Bull.,* 1096, 40pp

Hewlett, J.D., & Hibbert, A.R., 1967, Factors affecting the response of small watersheds to precipitation in humid areas. *International Symposium on Forest Hydrology Proc.,* 275-290

Hewlett, J.D., & Nutter, W.L., 1970, The varying source area of streamflow from upland basins. *Proc. Symposium on Interdisciplinary Aspects of Watershed Management,* 65-83

Kirkby, M.J., & Chorley, R.J., 1967, Throughflow, overland flow and erosion. *Bull. International Association Scientific Hydrology,* 12, 5-21

Lavee, H., 1973, *Relationship between the surface properties of the debris mantle and runoff yield in an extreme arid environment.* MSc. thesis. The Hebrew University. Mimeog. 71pp (Hebrew)

Levert, C., 1959, Some problems concerning the three dimensional location of a rain. *Kon. Nederl. Meteor. Inst., Wetenschappelijk Rapp.* WR 59-2, 39pp

Nutter, W.L., 1973, The role of soil water in the hydrologic behavior of upland basins. *Soil Science Society America Special Publication* 5, chapter 10, 14pp

Ragan, R.M., 1967, An experimental investigation of partial area contribution. *Bern Symposium International Association Scientific Hydrology,* 241-249

Rodda, J.C., 1967, The systematic error in rainfall measurements. *J. Institute Water Engineers,* 21, 173-177

Shachak, M., Chapman, E.A. & Steinberger, Y., 1976, Feeding, energy flow and soil turnover in the desert isopod, *Hemilepistus reaumuri. Oecologia,* 24, 57-69

Shanan, L., 1975, *Rainfall and runoff relationships in small watersheds in the Avdat region of the Negev desert highlands.* PhD thesis submitted to the Senate of the Hebrew University of Jerusalem.

Sharon, D., 1970, Topography - conditioned variations in rainfall as related to the runoff-contributing areas in a small watershed. *Israel J. Earth Sciences,* 19, 85-89

Weyman, D.R., 1974, Runoff processes, contributing area and streamflow in a small upland catchment. *Institute British Geographers Special Publication* 6, 33-43

Yair, A., 1972, Observations sur les effets d'un ruissellement dirigé selon la pente des interfluves dans une région semi-aride d'Israel. *Rev. de Géographie Physique et de Geologie Dynamique 2,* 14(5), 537-548

Yair, A., 1974, *Sources of runoff and sediment supplied by the slopes of a first order drainage basin in an arid environment.* Report of the Commission on Present Day Geomorphological Processes, Göttingen, 403-417

Yair, A., & Klein, M., 1973, The influence of surface properties on flow and erosion processes on debris-covered slopes in an arid area. *Catena,* 1, 1-18

6 TEMPORAL AND SPATIAL PATTERNS IN

EROSION AND FLUVIAL PROCESSES

*W. T. Dickinson & G. J. Wall

ABSTRACT

Temporal and spatial variability of erosion and fluvial sedimentation processes are examined for regions of Canada east of the Rocky Mountains. A spatial pattern has been determined for the rainfall erosion index, along with systematic differences in the monthly distribution of the index. As the seasonal pattern does not coincide with that of runoff and suspended sediment yield responses, a rainfall erosion index for spring conditions has been determined and mapped. Suspended sediment regime diagrams, dimensionless duration curves, and extreme value analysis reveal the runoff event orientation of fluvial sediment and its variability across the country. In light of the picture developed for temporal and spatial patterns, an erosion perspective is hypothesized.

INTRODUCTION

Traditionally the topics of land erosion and fluvial sedimentation have been treated in a "lumped" fashion, with rather coarse lumping in time and space models. For example, unit land erosion rates have been estimated from plot relationships and applied to large agricultural areas. Annual suspended sediment yields have been determined for large rivers and related to macroscale climate and physiography. Further, delivery ratios or transport have been used as an attempt to transcend the gap between soil loss estimates and annual sediment loads.

The lumped approaches have served well in introducing scientists to erosion and fluvial sedimentation systems. However, these approaches have been found wanting with regard to the determination of reliable prediction equations and have come up short as aids for the improvement of our understanding of the natural processes.

Recently the manner of approaching erosion and sedimentation has become more distributed. The notions of partial and dynamic contributing areas have freed some aspects of spatial lumping, and more detailed magnitude-

* W. T. Dickinson is Associate Professor of Hydrology, School of Engineering, Ontario Agricultural College, University of Guelph, Ontario, Canada
  G. J. Wall is a Soil Erosion Specialist, Canada Department of Agriculture, Guelph, Ontario, Canada

frequency and temporal analyses have afforded improved definition in time. With the advent of more distributed approaches have come renewed interest and refreshing hypotheses regarding erosion processes, their relative roles, and the mechanics of transport from source to stream channel.

The purpose of this paper is to utilize a distributed watershed perspective to consider the following questions. What has been resolved recently regarding temporal and spatial variability of erosion processes? Where are the knowledge gaps, and what approaches offer promise for closing such gaps?

PATTERNS OF EROSION POTENTIAL

Detailed erosion studies conducted during the past 30 years have identified several parameters that have been correlated strongly to soil loss (Musgrave 1947; Wischmeier and Smith 1965). These parameters have included a rainfall or runoff variable, a soil erodibility factor, slope gradient and length terms, and a soil cover variable. Wischmeier and Smith (1958) observed in plot studies that a rainfall energy interaction term (kinetic energy x intensity) provided a high correlation with soil loss values. The average annual value of this rainfall term has become known as the Rainfall Erosion Index (Wischmeier and Smith 1965).

When all the other factors affecting soil loss are held constant, the rainfall erosion index can provide a relative spatial mapping of rainfall-induced erosion potential. It has been found to range in the United States from a low of less than 50 to a high of approximately 600 (Wischmeier and smith 1965). Wall *et al* (1976) have recently computed rainfall erosion indices for Canadian sites east of the Rocky Mountains. Highest values were found in southwestern Ontario and the eastern extremities of the Maritime provinces (Figure 1). All values determined from Canadian data, ranging from less than 50 to 120, correspond well to the USA mapping and are at the low end of the range reported for the United States.

The rainfall erosion index can also be used to illustrate temporal patterns of rainfall-induced erosion potential. Figure 2 reveals the monthly distribution of the index for four representative Canadian stations (Wall *et al* 1976). While the distributions differ from station to station, a common feature of the temporal patterns is the high proportion of the rainfall erosion index contributed during the summer months of June through August. This seasonal pattern reflects the occurrence of convective storm activity during the summer months, and coincides with a pronounced seasonal pattern in extreme rainfall amounts observed by Dickinson (1976).

Erosion research conducted on plots both in the United States (USDA-ARS, Coschocton, Ohio) and in Canada (Ketcheson *er al* 1973) indicates that erosion from bare soils occurs primarily in the growing season when rainfall intensities are high and rain drops are larger,exhibiting

AVERAGE ANNUAL VALUES OF RAINFALL FACTOR (R) FOR EASTERN CANADA

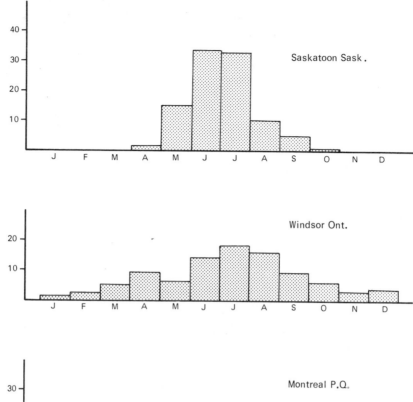

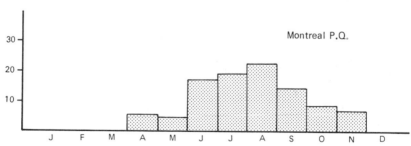

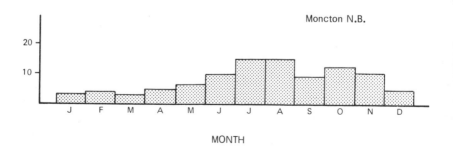

Figure 2   Seasonal distribution of erosion index values
            for Canadian stations

high values of kinetic energy. Further, Van Vliet *et al* (1976) have conducted plot studies in Guelph to determine erosion losses associated with snowmelt during winter and spring periods. The results of these studies have indicated that less than 5 per cent of the annual soil loss has been associated with snowmelt events. The combined results of the various plot studies reveal that the temporal pattern of soil loss from plot-size areas east of the Rocky Mountains is similar to the pattern of rainfall-induced erosion potential.

How do the above-noted patterns compare with that associated with stream runoff? The temporal and spatial variability of runoff has been considered by hydrologists for many years as a means of designing flood control structures and programs. Figure 3 illustrates the seasonal distribution of runoff volumes that may be expected in the Great Lakes Basin. Runoff volumes are largest in the late winter and spring periods, decline through the summer, and begin to increase again in the autumn. Since runoff has been identified as an important component of the soil erosion process, it seems paradoxical that the time of occurrence of maximum runoff differs so markedly from that of maximum soil loss. It is apparent that the temporal pattern of the rainfall erosion index does not reflect the pattern of watershed runoff response. A similar observation has been made by McGuinness *et al* (1971) for Northern Ohio watersheds.

Whereas the seasonal pattern of the rainfall erosion index does not coincide with the watershed runoff response pattern, and whereas the annual rainfall erosion index is determined largely by summer rainfall characteristics, there is reason to believe than an index of rainfall occurring during the spring period of maximum runoff generation could yield better information regarding the transport of eroded soil. Figure 4 has been prepared to show the spatial distribution of rainfall erosion potential determined only for periods of peak water runoff. The rainfall erosion index for spring conditions ranges from 4 to 33; three relatively distinct regions may be identified with Southwestern Ontario and Maritime areas exhibiting the highest values and Alberta the lowest. A large region encompassing

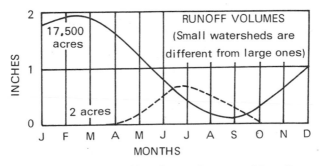

Figure 3   Seasonal distribution of run-off volumes in the Great Lakes basin

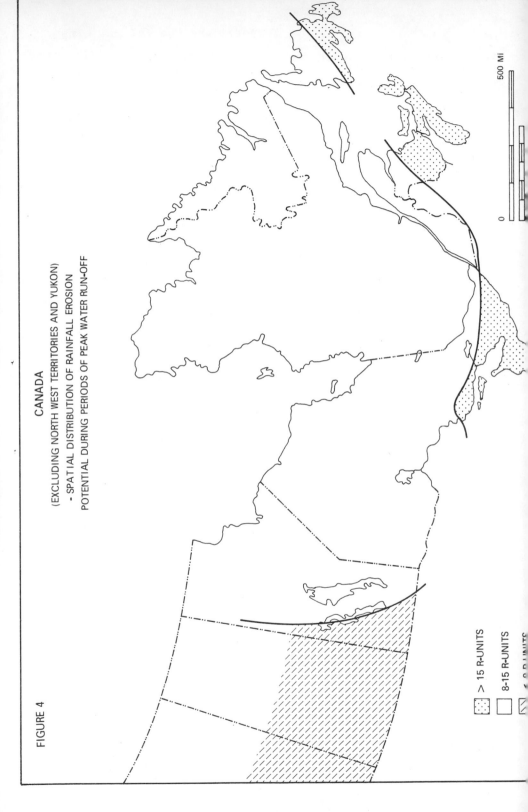

FIGURE 4

CANADA
(EXCLUDING NORTH WEST TERRITORIES AND YUKON)
- SPATIAL DISTRIBUTION OF RAINFALL EROSION
POTENTIAL DURING PERIODS OF PEAK WATER RUN-OFF

> 15 R-UNITS

8-15 R-UNITS

< 8 R-UNITS

500 Mi

0

much of Manitoba, Northern Ontario, Quebec, and Western
Newfoundland all exhibit an intermediate level of rainfall
erosion potential during the spring runoff period.

The temporal and spatial patterns of rainfall erosion
indices in Canada, when viewed in relation to runoff
patterns, promote the following observations. First, watershed runoff response is not reflected by either the seasonal
distribution of rainfall energy or an annual index developed
from such a distribution. Secondly, although the seasonal
distribution of rainfall energy suggests a potential for
soil erosion throughout the year peaking during the summer,
sediment transport to streams corresponds more closely to
the runoff pattern with peaks during spring snowmelt and
runoff period.

SEASONAL PATTERNS OF FLUVIAL SEDIMENTATION

The suspended sediment load transported in a stream is an
integrating variable which reflects the downstream contribution of upland surface erosion. It integrates in time
and space the processes of erosion and overland transport
within the watershed upstream of the stream station.
Although this integration masks to a considerable extent the
individual processes, consideration of the temporal and
spatial variability of suspended sediment load affords considerable insight into the upstream system, including the
efficiency with which a catchment transports its eroded
materials.

Annual suspended sediment load hydrographs clearly
reveal the time and magnitude of suspended sediment events,
and numerous examples of such hydrographs are now available
for Canadian rivers (eg. Robinson 1972;   Stichling 1973;
Dickinson *et al* 1975). The seasonal pattern is also
illustrated in a so-called regime diagram - a plotting of
the total suspended sediment load carried during each month
expressed as a percentage of the total load carried by the
river during the period of record. Such regime diagrams,
illustrated by Gregory and Walling (1973), are particularly
useful for comparing the regimes of different rivers, eg.
single-peaked vs multiple-peaked patterns. As it is
inevitably simpler to hypothesize a model (ie. develop a
concept regarding the significant processes and their
linkages) for single peak situations, a regime diagram provides a useful tool for identification of the complexity of
the watershed system being considered.

The apparent range of regime diagrams for Canada is
presented in Figure 5. It is evident from this figure that
the mountain and prairie suspended sediment regimes are
single-peaked, while there is an increasing tendency towards
a minor secondary peak as one moves eastward to Ontario and
the Maritimes. These sketches further reveal that: i) the
bulk of the annual load is transported in a very short
period of time;   ii) the time of major transport and peak
events does not correspond to the pattern of extremal
rainfall;   iii) suspended sediment peaks are strongly linked
to major runoff response, which occurs primarily as a
result of snowmelt, rain on snow, and rain falling on wet

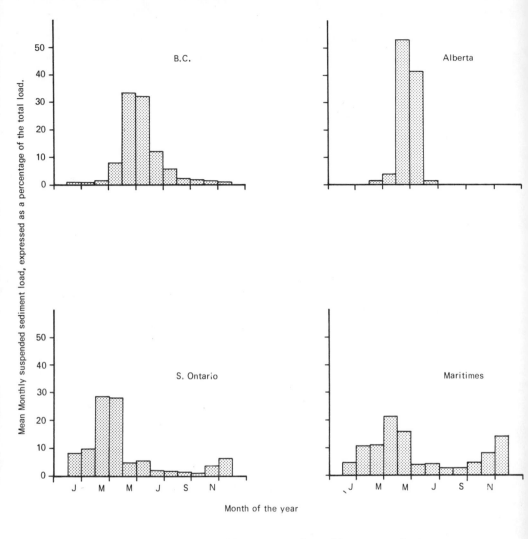

<u>Figure 5</u>   Suspended sediment regime diagrams for
selected regions in Canada

conditions prior to the summer drying season;   and iv)
erosion occurring as a result of summer rainfall seldom
reaches the stream and contributes negligibly to the annual
suspended sediment load.

It should be noted that size of watershed also has an
effect on the shape of the regime diagram of a river – the
smaller the basin, the sharper and more concentrated the
peak(s).   The examples presented in Figure 5 are for
moderate-sized basins.

FREQUENCY DISTRIBUTIONS

Long term variations in suspended loads are revealed in a
duration curve analysis of the data.   Examples of this
approach being used in sediment research have been discussed recently by Robinson (1972), Gregory and Walling
(1973), and Dickinson and Wall (1976).   Duration curves for
various rivers may be readily compared by means of a
dimensionless magnitude axis, yielding relationships
between the percentage of suspended sediment contributed by
suspended loads greater than or equal to selected values
and the percentage time that these selected values are
equalled or exceeded.   Examples of dimensionless duration
curves are presented for Canadian regions in Figure 6.

A number of observations may be made with regard to
Figure 6. i) Suspended sediment loads for all regions in
Canada have highly skewed frequency distributions,
suggesting that the transport of such loads is a discrete
process dependent upon extreme events - in fact, this
pattern is reflected in many parts of the world. For such
distributions, the mean loads (equalled or exceeded less
than 20 percent of the time) do not afford good indicators
of central tendency of the data.   Median loads are better
indicators of central tendency, while mean loads reflect
the range and total annual yield.   ii) For the preparation
of Figure 6, duration curves were determined for all streams
in Canada on which there were Water Survey data.   An
analysis of variance revealed that the regions in which the
rivers were situated (ie. the regions noted on the figure)
accounted for a significant amount of the variability
amongst the curves in the range of the percentage time axis
from 0 to 30 percent.   A median curve was selected to
represent each region considered.   iii) A comparison of the
regional curves reveals a systematic variation, but one that
is not immediately comparable to the earlier regime analysis.
For the dimensionless duration curves, the British Columbia
curve is similar to Ontario and the Maritimes,while the
Prairie and Alberta Mountain curves are set apart.   It would
appear that the dryer climatic region of the country
exhibits the most highly skewed distributions ie. the
percentage of time in which eroded material is transported
out of the basins is smallest for the more arid areas. For
the more humid regions - west of the Rockies and east of
the Prairies - whether or not the suspended sediment
regimes are single- or multiple-peaked, transport of
material occurs during a greater percentage of the time.

SEDIMENT EXTREMES

Where a sufficiently long period of record is available,
analysis can be extended to individual storms and daily
flows to evaluate their relative importance in contributing
suspended loads.   Archer (1960) noted that severe storms
may flush as much or more material in a few days than moves
through the system during an average year.   Wolman and
Miller (1960) did not note extreme variability for large
rivers but observed that as river flow variability increased,

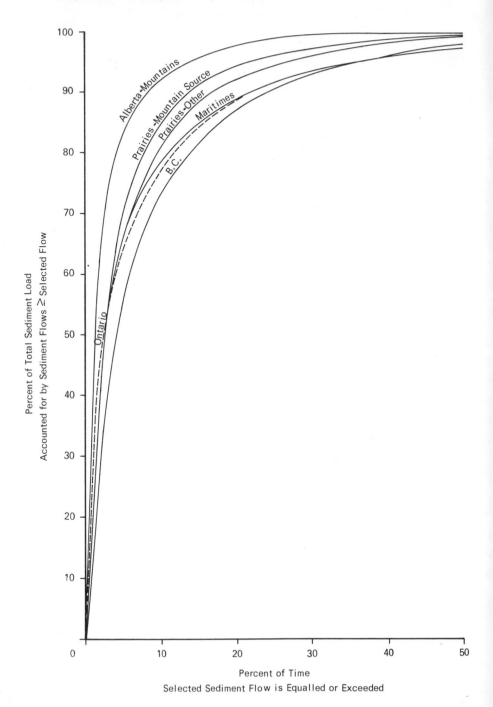

<u>Figure 6</u> Dimensionless suspended sediment duration
curves for regions in Canada

a larger percentage of the total load was carried by less frequent flows. Priest (1965) noted that the yield of storms with return periods greater than two years could represent from .3 to 45 percent of the total yield. Robinson (1972) found that for Canadian rivers 90 percent of the suspended sediment has been transported by daily flows which have recurred at least twice a year.

The lack of sufficiently long periods of suspended sediment record still prevents extensive or conclusive extremal analysis. However, an example extreme value plot is offered for the Thames River at Ingersoll (Figure 7).

In general, the slope of a sediment extreme value plot might be expected to be greater than the slope of the water flow extremes for a selected river station. However, an initial analysis of Canadian river data suggests that the relative slopes of suspended sediment and flow extremes can vary considerably (Figure 8). In some cases the slopes are very different, while in other cases they are quite close. Whether this variability is due to real differences between rivers, or is due to sampling error (largely attributable to short periods of record), the picture cannot be ascertained at this time. When a few more years of record become available, relationships such as those in Figure 8 will aid considerably in the determination of the relative efficiencies of drainage basins for transporting eroded material during peak events.

ANNUAL YIELDS

Apart from the spatial variability of temporal patterns noted above, mappings of annual suspended sediment yields have been presented for Canada by Fournier (1960), Strackhov (1967) and Stichling (1973). The relative yields for the regions identified in Figure 6 are presented in Table 1. These values coincide with those in the literature, but the spatial pattern is again somewhat different from the distributions of regimes and of dimensionless duration curves. It would appear that the temporal characteristics are affected primarily by climatic factors, while magnitudinal or scale characteristics are linked further to physiography.

Table 1   Regional mean annual sediment loads for Canada

Region	Load (metric tons/ km^2 /year)
British Columbia	90
Alberta - Mountains	75
Prairies - Mountain Source	50
Prairies - Other	25
Ontario (Southern)	15
Maritimes	25

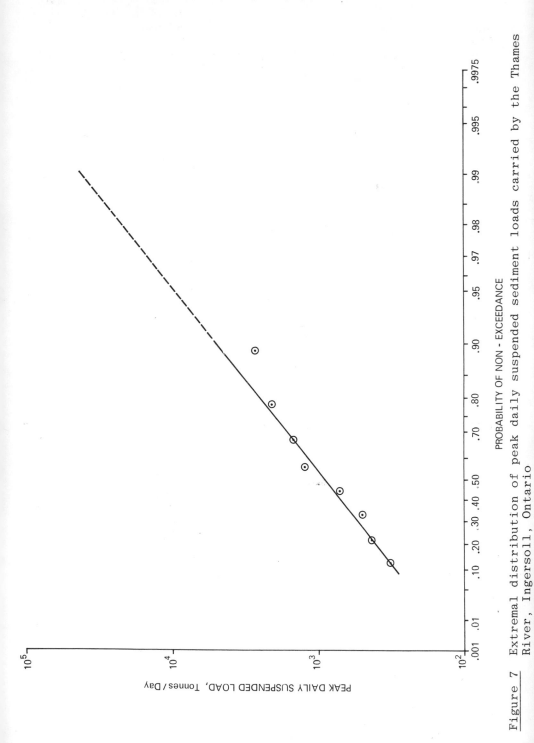

Figure 7 Extremal distribution of peak daily suspended sediment loads carried by the Thames River, Ingersoll, Ontario

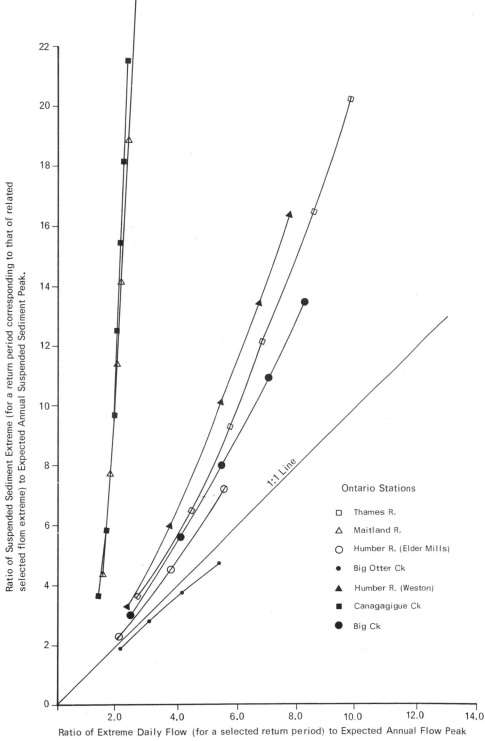

Figure 8 Relationships between suspended sediment and flow
extremes for selected river stations

AN EROSION PERSPECTIVE

The patterns of temporal and spatial variability associated with soil loss and suspended sediment suggest the following hypothetical perspective. i) Land erosion is widespread in the spring of the year due to snowmelt, rain on snow, and rain falling on bare wet soil. Although point erosion rates during this period of the year may not be the highest for the year, basins exhibit at this time a large capacity to transport eroded material to streams and further downstream. The effective delivery ratio during the spring may be expected to be relatively high, and the highest of the year. ii) Land erosion is not widespread during the summer ie. during the period after the evaporative demand of the local climate has dried soil conditions. This drying occurs very rapidly in most areas of the Canadian landscape. Although very localized high erosion rates may occur - often on a plot scale, seldom on a field scale, and very rarely on a watershed scale - the effective delivery ratio appears to be close to zero during the summer season.

In the light of this perspective of erosion processes, a number of topics yet requiring attention and further research become apparent.

i) Delineation of source areas: Although erosion source areas and sediment contributing areas are being investigated, there is a need to identify the nature of the areas, their variability in time, and other watershed variables indicative of source areas. Neither the location of summer erosion "pockets" nor the relationship of such areas to spring contributing areas has received much attention. The summer "pockets" may well be a subset of spring sources or may be an independent group which never contribute material to stream sediments. Further, summer eroded material may become revegetated and reincorporated and hence seldom be available for subsequent spring movement. The identification of seasonal erosion and sediment source areas and their behaviour could be profitably considered in relation to hydrologic contributing area models.

ii) Determination of transport rates: The statements above regarding the possible seasonal variability of delivery ratios, although intuitively reasonable at this point, are quite hypothetical. Studies to date have not addressed this topic. There is a need for research with regard to not only the variability in time and space of delivery ratios, but also the possible relationships between such variability and that associated with the hydrologic runoff coefficients. Further, even when material is transported through the watershed system, the actual transport distances from field to stream to downstream and their frequency of occurrence are not well characterized.

When results become available regarding these topics, and are cast in the setting of a realistic erosion perspective or concept, then a meaningful model of erosion and fluvial sedimentary processes will be developed. Such a model

will not only clarify understanding of the system but also
provide a useful means for attacking such practical
problems as the design of remedial measures to control the
movement of sediment and associated contaminants.

ACKNOWLEDGEMENTS

The authors wish to acknowledge the analytical assistance
and useful suggestions of Mr. Q. Khan, Mr. A. Matheson and
Mr. L. J. P. Van Vliet. The study has been carried out as
part of the efforts of i) the International Reference
Group on Great Lakes Pollution from Land Use Activities,
an organization of the International Joint Commission, and
ii) OMAF Research Program 38 - Agriculture and Water
Quality. Funding has been provided through Agriculture
Canada and the Ontario Ministry of Agriculture and Food.
Findings and conclusions are those of the authors and do
not necessarily reflect the views of the Reference Group,
its recommendations to the Commission, or of OMAF.

REFERENCES

Archer, R.J., 1960, Sediment discharge of Ohio streams
        during floods of January-February 1959. *Ohio Dept.
        Natural Resources, Division of Water, Columbus, Ohio.*

Dickinson, W.T., 1976, Seasonal variability of rainfall
        extremes. *Atmosphere,* 14(4), 288-296

Dickinson, W.T., Scott, A. & Wall, G., 1975, Fluvial
        sedimentation in Southern Ontario. *Canadian J. Earth
        Sciences,* 12(11), 1813-1819

Dickinson, W.T. & Wall, G.J., 1976, Temporal pattern of
        erosion and fluvial sedimentation in the Great Lakes
        Basin. *Geoscience Canada,* 3(3), 158-163

Dickinson, W.T. & Wall, G.J., 1976, The influence of land
        use on erosion and transfer processes of sediment.
        *IJC Workshop on Fluvial Transport of Sediment-
        Associated Nutrients and Contaminants. Kitchener,
        Ontario.*

Fournier, F., 1969, Transports solides effectivés par les
        cours d'eau. *IASH Bulletin,* 14(3), 7-47

Gregory, K.J. & Walling, D.E., 1973, *Drainage Basin Forms
        and Process.* Fletcher & Son Ltd., Norwich

Ketcheson, J.W., Dickinson, W.T. & Chisholm, P.S., 1973,
        Potential contributions of sediment from agricultural
        land. *Proc. 9th Canadian Hydrology Symposium,
        Edmonton.*

McGuiness, J.L., Harrold, L.L. & Edwards, W.M., 1971,
        Relation of rainfall energy streamflow to sediment
        yield from small and large watershed. *J. Soil and
        Water Conservation,* 26, 233-235

Musgrave, G.W., 1947, The quantitative evaluation of
        factors in water erosion, a first approximation.
        J. Soil and Water Conservation, 2, 133-138

6. Temporal and spatial patterns

Piest, R.F., 1965, The role of the large storm as a sediment contributor. *Proc. 1963 Federal Interagency Conference on Sedimentation. US Department of Agriculture Miscellaneous Publication* 970, 97-108

Robinson, M.W., 1972, Sediment transport in Canadian streams: a study in measurement of erosion rates, magnitude and frequency of flow, sediment yields, and some environmental factors. MSc Thesis, Department of Geography, Carleton University, Ottawa, Canada

Stichling, W., 1973, Sediment loads in Canadian Rivers. *Proc. 9th Canadian Hydrology Symposium, Edmonton.* 39-72

Strackhov, N.M., 1967, *Principles of lithogenesis.* Translation by J. P. Fitzsimmons, S.I.Tomkieff and J.E.Hemingway. New York

Tywoniuk, N. & Cashman, M.A., 1973, Sediment distribution in river cross-sections. *Proc. 9th Canadian Hydrology Symposium, Edmonton.* 73-95

Van Vliet, L.J.P. & Wall, G.J., 1976, Soil loss from winter runoff events in southern Ontario. *Abstracts, Canadian Society Soil Science, Annual Meeting, Halifax*

Wall, G.J., van Vliet, L.J.P. & Dickinson, W.T., 1976, The universal soil loss equation - method of predicting soil loss in Canada. *Abstracts, Canadian Society Soil Science, Annual Meeting, Halifax*

Walling, D.E., 1976, Erosion processes, sediment sources, and sediment yields. *IJC Workshop on Fluvial Transport of Sediment-Associated Nutrients and Contaminants. Kitchener, Ontario*

Wischmeier, W.H. & Smith, D.D., 1958, Rainfall energy and its relationship to soil loss. *American Geophysical Union Trans.,* 39(2), 285

Wischmeier, W.H. & Smith, D.D., 1965, Predicting rainfall erosion losses from cropland east of the Rocky Mountains. *US Department of Agriculture Handbook* 282

Wolman, M.G. & Miller, J.P., 1960, Magnitude and frequency of forces in geomorphic processes. *J. Geology,* 68, 54-74

7 AN APPROACH TO THE PREDICTION OF SUSPENDED SEDIMENT RATING CURVES

*W.F. Rannie

ABSTRACT

Because existing formulae for predicting suspended load transport cannot be applied to streams in which wash load comprises a significant proportion of the total, suspended load estimates for most streams are normally obtained from sediment rating curves of the form $G_S = kQ^j$ where G_S = sediment discharge and Q water discharge. There is at present no method of predicting the parameters of sediment rating curves for ungauged watersheds; nor are there standards by which rating curves from different basins may be compared.

In this study it is shown that the constant k and exponent j for rating curves from a wide variety of erosional environments are related by the empirical equation $j = 1.581 - 0.155 \log k$ (in cfs - tons per day units). This relationship implies that rating curves may be modelled by a family of lines which intersect at a common point and which have the general form

$$G_S = c \, Z \, Q^{1.581 - 0.155 \log c - 0.155 \log Z}$$

where $k = c \, Z$ and $j = 1.581 - 0.155 \log k$. The parameters c and Z are assumed to reflect the physical characteristics of the watershed which govern the erosion-sediment transport system.

Multiple regression of $\log k$ and a set of morpho-climatic variables for fifty widely distributed watersheds from the United States and southern Ontario yielded the following best-estimates for c and Z:

$$c = 118.0; \quad z = q^{-2.304} H^{-1.209} A^{-0.676} \qquad (r = 0.85)$$

$$c = 239.0; \quad z = q^{-2.381} \bar{X}^{-1.511} A^{-0.677} \qquad (r = 0.86)$$

where q = mean annual runoff (cfs/mi^2), H = maximum relief (ft), $\bar{X}$ = mean relief (ft) and A = watershed area (mi^2). No substantial improvement resulted when drainage density, channel slope and basin hypsometry were included in the variable set.

The probable errors in the prediction of sediment transport from these rating functions at $Q = 1.0$ cfs (ie. where $G_S = k$) are 285% and 274%. Because the exponent and constant are inversely related, however, these errors

*Department of Geography, University of Winnipeg, Winnipeg, Manitoba, Canada

7. Prediction of rating curves

*normally become smaller as discharge increases. Conse-
quently, for the fifty study basins, the average errors in
the prediction of maximum suspended load decreased to 143%
and 147%. Examination of the residuals suggests that a
large part of the unexplained variance is contributed by
basins with short records and by basins in the Mediterranean
climatic zone. It is concluded that the general rating
functions proposed here predict long-term average suspended
load transport for basins with 'continental' climatic
regimes at least as accurately as other available methods
and require only extensive, easily-measured basin properties.*

*Further research should concentrate upon the testing
of a more intensive set of variables for the prediction of
k and upon the development of relationships based upon broad
morphoclimatic regions. In addition to their predictive
value, such relationships would provide a standard against
which rating curves in general may be compared.*

INTRODUCTION

Most suspended load formulae are based upon the assumptions
(a) that the amount of sediment being transported is a
function of the equilibrium between the transport capability
of the flowing water and the availability of suitable
material on the stream bed, and (b) that the sediment transport rate is ultimately predictable from the hydraulic
properties of the flowing water, the stream channel and the
bed material. It follows, then, that such formulae can
predict only the quantities of sediment derived from the
stream bed (bed material load) and must exclude wash load;
in effect, natural streams are treated as flumes. In most
streams, however, wash load originating from sheet, rill,
and gulley erosion constitutes more than 80% of the total
suspended load (Einstein 1964). The problem this poses
for sediment transport estimation was recognized by Shen
(1972):

> If the hypothesis is true that large rivers
> transport a great deal of wash load ... and only
> a small amount of bed material load, the con-
> sequence is rather alarming. That means one must
> estimate the total sediment transport rate in
> these large rivers mainly from their upstream
> wash load supplies rather than from their
> capabilities to transport. (p 14)

As a result, most sediment measurement programs throughout the world have adopted the sediment rating curve as a
means of estimating suspended sediment transport from flow
data. Such curves are normally found to assume a power
relationship (Equations 1 and 2), although departure from
this form may occur in the highest and lowest discharge
ranges.

$$G_S = k \ Q^j \qquad (1)$$

$$\log G_S = \log k + j \log Q \qquad (2)$$

where G_S = suspended load discharge (MT^{-1}); Q = water discharge (L^3T^{-1}); k, j = coefficient and exponent respectively for a particular gauging site and measurement system.

The exponent j of the rating curve reflects the average rate at which sediment transport increases for a given increase in water discharge. If water discharge is considered to be an index of the intensity of fluvial processes operating within a watershed-channel system, j is a measure of the rate at which a change in process level is converted to additional geomorphic work. Thus, it might be used as a measure of the metabolism of the system, such that low exponents indicate relative insensitivity to changes in the level of the process and high exponents indicate a relatively sensitive system.

Because the suspended load of most streams is derived mainly from processes which operate throughout the watershed and is imposed upon the stream channel, Leopold and Maddock (1953) reasoned that the suspended sediment concentration is an independent variable to which other elements of the hydraulic geometry must adjust:

The at-a-station ... relations between discharge and suspended load ... may be considered essentially independent of the channel system, but are functions of the drainage basin. The relations of width, depth and velocity to discharge tend to be adjusted to conform to the load-discharge function.   (p 28)

Thus the rate of increase in sediment transport with water discharge provides, at least in part, a link between watershed and channel processes. While the nature of this linkage may be investigated by studying the channel, an explanation for variations in j must be sought in the watershed. This was recognized by Bogardi (1961) who found a good relationship between the rating curve exponents for stations on twelve Hungarian rivers and four hydrologic variables: average width of watershed (area/basin length), mean discharge, ratio of highest to lowest discharges, and mean discharge per unit area. Although Bogardi's relationships cannot be extrapolated to other environments, his study is important as it appears to be the only one which has attempted to explain variations in the rating curve exponent in terms of watershed characteristics.

The coefficient k of the rating curve has received little attention, perhaps because it is less expressive of the dynamic qualities of the fluvial system. Nonetheless, k represents the transport rate when Q = 1.0 and for a given exponent, it sets the level of the process in absolute terms. It might be expected, therefore, that k is also related to the characteristics of the system it describes. Bauer and Tille (1967) found a good relationship (r = -0.836) between k and mean annual discharge for eleven rating curves in the German Democratic Republic but did not extend their analysis to examine the effects on k of other watershed characteristics.

7. Prediction of rating curves

Despite the almost universal acceptance of the suspended sediment rating curve as a hydrologic and geomorphologic tool, rating curves themselves have received surprisingly little attention. Apart from the geographically-restricted studies of Bauer and Tille and Bogardi referred to above, no general standards exist whereby rating curves from watersheds of differing character may be compared; consequently, rating curves are usually reported in the literature in isolation from those which have been determined elsewhere. Nor have norms been established which might indicate whether a particular rating curve is unusual for its geohydrologic environment and thus worthy of further investigation. Furthermore, because the rating curve remains the only practical method of relating suspended load transport to water discharge, a method of predicting the position of the rating curve from environmental variables would have considerable practical value as a means of obtaining reconnaissance estimates of sediment transport for streams which have no sediment measurement programme. It is the intention of this paper, then, to demonstrate that the parameters of sediment rating curves are susceptible to prediction.

A MODEL FOR A GENERAL SEDIMENT RATING CURVE

If both j and k are related to the physical character of the system as the foregoing discussion suggests, it is reasonable to suppose that they may themselves be related and may covary in a consistent fashion. To test this, the writer assembled all available rating curves from the sediment transport literature. Data for sixty-nine stations on forty-five rivers throughout the world indicate a clear logarithmic relationship between j and k ($r = -0.913$), such that

$$j = 1.516 - 0.168 \log k \tag{3}$$

in tons per day-cfs units (Figure 1A). In addition, one or both of the parameters for thirty-one rating curves were extracted from graphs and while these data were not used to define Equation 3 because of the inaccuracy inherent in the interpolation, they appear to follow the same relationship (Figure 1B).

When the slope and intercept of a set of linear equations are themselves linearly related (as in Equation 3), the equations form a family of lines which intersect at a common point and which can be modelled by a function of the form

$$Y = c + dX + eZ + g(XZ) \tag{4}$$

When the variables X, Y and Z, and the constant c are expressed in logarithmic form, Equation 4 may be restated:

$$\log Y = \log c + d(\log X) + e(\log Z) + g(\log X \log Z) \tag{5}$$

With $Y = G_s$ (sediment discharge) and $X = Q$ (water discharge), Equation 5 has the form of a sediment rating function:

$$G_s = cZ^e \, Q^{d + g \, (\log Z)} \tag{6}$$

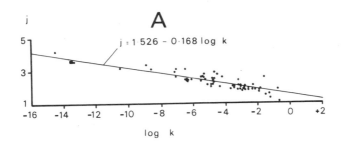

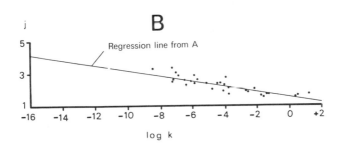

<u>Figure 1</u> Log k vs j for rating curves reported in
literature. (A) where equations are
given; (B) extrapolated from graphs.

where $k = cZ^e$ and $j = d + g (\log Z)$. Combining Equations
3 and 6 and arbitrarily assuming $e = 1.0$ produces the
simplified form

$$G_s = cZQ^{1.516 - 0.168 \log c - 0.168 \log Z} \qquad (7)$$

If it is assumed that Z is a multivariate factor
incorporating the combined effects of all variables which
control the watershed erosion-sediment transport system,
then

$$k = f(Z) = f (X_1, X_2, X_3 \ldots\ldots\ldots X_n)$$

where X_i are variables which characterize the system. An
estimate of $f(Z)$ and thus of the entire general rating
function may be obtained by multiple regression of k and a
set of watershed variables for a sample of basins.

METHODOLOGY

Within a general objective of including the broadest possible
range of geomorphic environments, the drainage basins used
to evaluate $k = f(Z) = f(X_i)$ were selected according to the
following criteria:
a) availability of daily water and sediment discharge
   records;
b) availability of large-scale (1:24,000 or 1:25,000)
   topographic maps;
c) absence of significant lake or reservoir area within the
   basin; and

d) basin area of less than 300 square miles. The restriction on the size of the basins was dictated in part by the logistical problems imposed by the map scale, and in part by the desire to minimize the environmental discontinuities which might occur within an individual basin. The upper limit of 300 square miles was somewhat arbitrary, however, and one significantly larger basin, the Rio Puerco (420 square miles), was included because it represented unusually high transport rates in a semi-arid environment.

Records fulfilling the stated requirements were obtained for forty-seven basins in the United States and three basins in Southern Ontario, Canada (Table 1, Figure 2).

A sediment rating curve in tons per day-cfs units was obtained for each stream by sampling the published daily water and sediment discharge records. In order to ensure that the entire length of record was utilized, and that the full range of flows was represented, the record was stratified into 10-day periods and the highest water discharge and its associated sediment load was selected from each period. This procedure proved more satisfactory than random or stratified random schemes because it ensured that the upper and most important part of the rating curve was well-defined. In fact, it guaranteed the inclusion of the highest mean daily flow during the period of record. The only departures from this procedure were for ephemeral streams in which flow occurred on a relatively small number of days each year, and for streams with records of less than three years. In the former case, all flow values were used; in the latter, the time period for sampling was reduced to five days in order to obtain a larger sample.

The distribution of water discharges obtained from this sampling procedure was skewed toward the lower flows and consequently the contribution of each flow level to the position of the best-fit line through the sample points was inversely proportional to its magnitude. In order to reduce the excessive influence of low flows on the rating curve position, the flow range was subdivided geometrically into 10-12 intervals. The average logarithm for all sampled water and sediment discharges was computed for each interval and the best-fit line was computed from these averages. This procedure had the effect of weighting the individual sample points in direct proportion to the absolute discharge; ie. small flows contributed less to the final determination of the rating line than did large flows.

The variables which were used to describe each watershed and to predict k were selected according to the following criteria: (i) they should primarily reflect qualities of the watershed rather than local properties of the river channel in the vicinity of the gauging station; (ii) they should be simply and objectively determined; (iii) they should be universal in the sense that they represent qualities common to all watersheds; and (iv) they should be measurable on an open scale.

The principal effect of criteria (iii) and (iv) was to eliminate from consideration variables which describe, for

<u>Figure 2</u>   Location of sample basins

example, forest cover, area of lakes, land use, etc.
Although such variables may be of considerable local or
regional importance, their inclusion would have geogra-
phically restricted the usefulness of the results.

The following watershed attributes were selected as
being central to any definition of the system as well as
fulfilling the requirements stated above: area, relief,
relief distribution or hypsometry, mean annual run-off,
drainage density, and channel slope.

Relief was expressed in three ways - maximum or
absolute relief, standard deviation of relief, and mean
relief. The latter two measures were obtained by sampling
elevation at several hundred points throughout the basin
and calculating the mean and standard deviation from the
first and second moments of the distribution. Hypsometry
was expressed by the skewness and kurtosis of the distribution, obtained from the third and fourth moments
respectively, as was suggested by Tanner (1959, 1960) and
Evans (1972).

The blue-line method of measuring drainage density
was adopted as an expedient because the application of
other more accurate methods is impractical for basins of
several hundred square miles. In any case, measurement of
true drainage density at the scale at which hillslope processes operate is virtually impossible and all methods
express relative drainage density only. Thus, when drainage

7. Prediction of rating curves

Table 1 List of basins selected for study

No	Streams and gauging station
1	Scantic River at Broad Brook, Connecticut
2	Kayaderosseras Creek near West Milton, New York
3	Stony Brook at Princeton, New Jersey
4	Brandywine Creek at Chadd's Ford, Pennsylvania
5	South Branch Raritan River at Stanton, New Jersey
6	Perkiomen Creek at Graterford, Pennsylvania
7	Salem Fork at Salem, West Virginia
8	Hazel River at Rixeyville, Virginia
9	Hudson Creek near Boswell's Tavern, Virginia
10	Plum Creek at Waterford, Kentucky
11	West Branch Susquehannah River at Bower, Pennsylvania
12	North Fork Massie Creek at Cedarville, Ohio
13	South Fork Massie Creek near Cedarville, Ohio
14	Corey Creek near Mainesburg, Pennsylvania
15	Elk Run near Mainesburg, Pennsylvania
16	Bixler Run near Loysville, Pennsylvania
17	Todd Fork near Roachester, Ohio
18	Marsh Creek at Blanchard, Pennsylvania
19	Tygarts Creek near Greenup, Kentucky
20	Alum Creek at Columbus, Ohio
21	Mad River at Eagle City, Ohio
22	Thames River at Ingersoll, Ontario
23	Big Otter Creek near Vienna, Ontario
24	Big Creek near Walsingham, Ontario
25	Big Raccoon Creek near Fincastle, Indiana
26	Black Earth Creek at Black Earth, Wisconsin
27	Yellowstone River near Blanchardville, Wisconsin
28	Ralston Creek at Iowa City, Iowa
29	Paint Creek near Waterville, Iowa
30	Dry Fork Creek near Chulahoma, Mississippi
31	Cuffawa Creek at Chulahoma, Mississippi
32	Elm Fork Trinity River near Muenster, Texas
33	Pin Oak Creek near Hubbard, Texas
34	Vermillion Creek near Wamego, Kansas
35	Dry Creek near Curtis, Nebraska
36	Mitchell Creek above Harry Strunk Lake, Nebraska
37	Bridger Creek near Lysite, Wyoming
38	Fivemile Creek above Wyoming Canal, near Pavillion, Wyoming
39	North Clear Creek near Blackhawk, Colorado
40	North Fork Crazy Woman Creek, near Greub, Wyoming
41	Kiowa Creek at Kiowa, Colorado
42	Rio Puerco below Cabezon, New Mexico
43	Uvas Creek above Uvas Reservoir near Morgan Hill,California
44	Coyote Creek near Gilroy, California
45	Arroyo Seco near Greenfield, California
46	Napa River near St.Helena, California
47	Naciemento River near Bryson, California
48	South Fork Eel River near Branscomb, California
49	Willow Creek at Heppner, Oregon
50	Mill Creek below Blue Creek near Walla Walla, Washington

densities in different basins are being compared, considerations of the comparability of map scales and cartographic conventions are probably more important than the actual definition of what constitutes the drainage net. Since all but three of the basins used were mapped by the same agency (United States Geological Survey) at a uniform scale, the standards for inclusion of streams were consistent, and it was felt that the resulting drainage density data would at least indicate relative degree of channel development.

A channel slope measure was included because it is the only channel characteristic which can be obtained from maps. Since slope varies continuously along a channel, an average slope must be defined over a reach appropriate to the size of the stream: for this study, reach length was arbitrarily defined as $A^{0.5}$. Thus, the slope parameter represents the average gradient over a channel distance equal to $A^{0.5}$ miles upstream from the gauging station.

Basin area and mean annual runoff (mean annual discharge per unit area) were accepted as reported in the hydrologic records.

Summary statistics of these variables for all basins given in Table 2 indicate the diversity of environments contained within the sample.

Table 2 Summary statistics for all sample basins

Variable	Mean	Maximum	Minimum
A	115.54	420.0	3.01
H	1563.78	6600.0	82.0
$\bar{X}$	577.97	2211.6	44.6
s	299.57	1508.4	20.9
P	3.17	7.85	1.52
Sk	+0.27	+2.07	-1.94
q	0.900	3.941	0.013
D	2.05	4.06	0.97
S	0.005	0.032	0.000473

where A = basin area (square miles); H = maximum relief (feet); $\bar{X}$ = mean relief (feet); s = standard deviation of relief (feet); P = kurtosis of relief distribution; Sk = skewness of relief distribution; q = mean annual runoff (cfs per square mile); D = drainage density (miles per square mile); S = mean channel slope.

RESULTS

For the fifty sample basins, the relationship between j and log k was:

$$j = 1.581 - 0.155 \log k$$

with $r = -0.76$ and $s_e = 0.213$ (Figure 3)

The parameters of Equation 8 are not significantly different (at the 99% level) from those obtained from the analysis of rating curves reported in the literature (Equation 3). However, Equation 8 is accepted here because

7. Prediction of rating curves

the rating curves on which it is based were derived in a consistent fashion.

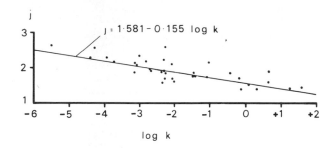

Figure 3 Log k vs j for sample basins

Substituting the parameters from Equation 8 into the general rating function (Equation 6) yields

$$G_S = cZ \ Q^{1.581 - 0.155 \log c - 0.155 \log Z} \tag{9}$$

in tons per day - cfs units.

Equations for predicting log k = f(log Z) = f(log X_i) were developed by step-wise multiple regression of kog k against the watershed variables. The constants and regression coefficients for three sets of equations (one set for each relief measure) are given in Table 3.

Basin hypsometry was also evaluated using untransformed skewness and kurtosis; however, the resulting equations have not been included in Table 3 because neither of these variables improved the correlation or the standard error of estimate. The use of relative relief terms (R_R = H/A $^{0.5}$; s/A$^{0.5}$) was similarly unproductive.

From Table 3, estimates of c and z for each of the three relief measures are:

Maximum Relief

$$c = 0.0178; \ Z = q^{-2.41} H^{-0.618} A^{-1.202} S^{-1.0} D^{1.556} \tag{10}$$

Standard Deviation of Relief

$$c = 0.0115; \ Z = q^{-2.446} S^{-0.671} A^{-1.19} S^{-0.937} D^{1.535} \tag{11}$$

Mean Relief

$$c = 0.291; \ Z = q^{-2.446} \bar{X}^{-1.038} A^{-1.038} S^{-0.723} D^{-.682} \tag{12}$$

These values may be inserted into the general rating function to provide equations which predict the position of the rating curves for unsampled basins. Because there is no substantial difference in the prediction accuracy of Equations 10, 11 and 12, and since the standard deviation of relief, s, is more tedious to compute than maximum relief, the discussion which follows will concentrate upon the mean and maximum relief equations.

158

(a) Prediction equations for log k using maximum relief (H)

Dependent variable	Equation constant	log q	log H	log A	log S	log D	R^2	Standard error (log units)
log k	-2.652	-2.000	-1.553				0.47	1.170
log k	+1.869	-2.290	-1.209				0.68	0.921
log k	+2.072	-2.304	-0.491	-0.676			0.72	0.867
log k	-1.538	-2.375	-0.491	-1.246	-0.982		0.74	0.847
log k	-1.750	-2.410	-0.616	-1.202	-1.000	+1.556	0.76	0.828

(b) Prediction equations for log k using standard deviation of relief (s)

Dependent variable	Equation constant	log q	log s	log A	log S	log D	R^2	Standard error (log units)
log k	-2.652	-2.000	-1.532				0.47	1.170
log k	+0.690	-2.300	-1.205				0.67	0.925
log k	+1.272	-2.320	-0.584	-0.726			0.73	0.858
log k	-1.530	-2.380	-0.671	-1.207	-0.892		0.74	0.841
log k	-1.938	-2.416	-0.671	-1.190	-0.937	+1.535	0.76	0.823

(c) Prediction equations for log k using mean relief ($\bar{X}$)

Dependent variable	Equation constant	log q	log $\bar{X}$	log A	log D	log S	R^2	Standard error (log units)
log k	-2.652	-2.000	-1.906				0.47	1.170
log k	+2.171	-2.382	-1.511				0.68	0.906
log k	+2.379	-2.381	-1.713	-0.677			0.73	0.849
log k	+2.265	-2.430	-1.038	-0.613	+1.752		0.75	0.825
log k	-0.536	-2.446	-1.038	-1.038	+1.682	-0.723	0.76	0.821

Table 3

7. Prediction of rating curves

Equations 10 and 12 both explain nearly 76% of the variance of log k, with standard errors of estimate of 0.83 and 0.82 log units. Taking percentage error as the difference between the actual and predicted values divided by the actual value, estimates of average sediment transport from Equations 10 and 12 for discharges in the vicinity of 1.0 cubic feet per second (when G_s = k) have a standard error of nearly 600%. The probable error (=0.6745 s_e)is 0.554 log units or 250%.

For discharges much greater than 1.0 cfs, the accuracy of the estimated sediment transport depends upon the errors associated with both the predicted position of the coefficient (log k) and the predicted slope of the rating curve. Four possible combinations of these errors exist (Figure 4):

(I) The coefficient is accurately predicted but the slope is in error. In this situation the predicted sediment transport will be accurate in the lower range of discharge (near 1.0 cfs) but will depart by an increasing amount as flow increases. The amount of this departure in log units is equal to the error in the estimate of slope times log Q.

(II) The coefficient is in error but the slope is accurately predicted. Here the predicted sediment transport will be in error by a constant amount equal to the error in the prediction of log k.

(III) Neither the coefficient nor the slope are accurately predicted and both errors are of the same sign; that is, both log k and j are overestimated or underestimated. This type of error is the most serious since the errors are compounded as Q increases.

(IV) Neither the coefficient nor the slope are accurately predicted but the errors are of opposite sign; that is, when log k is underestimated, j is overestimated and vice versa. In this case, the errors are conservative in that the error in the estimate of slope tends to 'correct' the error in log k as discharge increases.

Because of the power form of the rating curves, sediment transport is usually many magnitudes greater at high discharges than at low and for most purposes it is more important that the upper portion of the rating curve be estimated accurately. Errors of Type (IV) are clearly preferable since they correspond to increasing accuracy as discharge increases. Fortunately, because of the inverse relationship between log k and j (Equation 8), Type (IV) errors are the most probable result of the model proposed here. This is indicated on Figure 5 in which the residuals of log k (from Equation 12) are plotted against the errors in the estimate of j obtained from Equation 8 (using the predicted value of k). Increasingly positive and negative errors in the estimated log k are clearly associated with increasingly negative and positive errors respectively in j. Only seven of the fifty residual combinations are of the same sign (Type III error) and most of these are of small magnitude. An identical situation exists for the residuals from Equation 10.

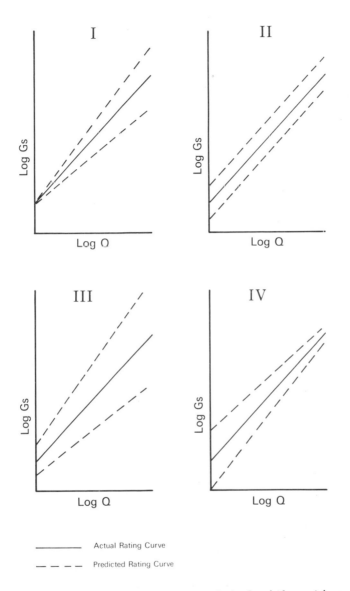

Figure 4   Types of errors associated with rating
curve prediction

As a further indication of the self-correcting nature
of the general rating function, sediment discharges computed
from Equation 12 for the maximum discharge group for each
basin were compared with those obtained from the actual
sediment rating curves. The standard deviation of the
differences between the actual and predicted values was
0.548 log units. Thus the probable error in predicted
sediment discharges is reduced from 250% at 1.0 cfs to

161

7. Prediction of rating curves

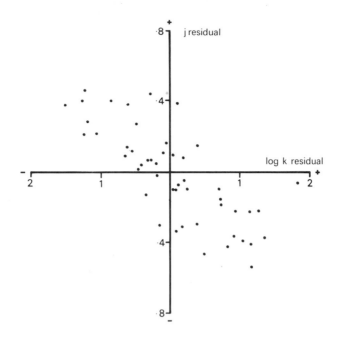

Figure 5   Errors in estimate of log k from Equation 12
vs errors in estimates of j from Equation 8

about 130% at peak flow.

While a probable error of this magnitude is not incon-
siderable, it must be viewed in the context of the errors
associated with other suspended sediment formulae. In a
review of the most widely used formulae, Vanoni (1975) concluded:

> ... one must assume that the probable error in
> sediment discharge calculations under the most
> favourable circumstances is large. Errors as
> large as 50% to 100% can be expected. When
> calculations are based on average values of
> slope, bed material characteristics, temperature
> and calculated flow depth, and velocity ...
> larger errors can be expected. (p 229)

All other formulae predict suspended bed material
transport only, despite the fact that in most streams wash
load is the dominant component. Furthermore, most of the
formulae require hydraulic data which are usually unavailable
for most streams and which are frequently more time-
consuming and expensive to obtain than a sediment sampling
programme. In contrast, the general rating functions
developed above are based upon measured data from an
extremely broad range of environmental conditions, they
include both suspended bed material and wash load, and they
can be estimated from readily-available data.   The fact that
they appear to be of comparable accuracy to more elaborate

formulae makes them an attractive means of obtaining reconnaissance estimates for ungauged streams.

Some of the factors which contribute to the unexplained variance of log k can be deduced from the residuals from Equations 10 and 12. Two patterns emerge: 1) consistently poor prediction of log k for California streams; 2) insufficient record length used to establish some of the sample rating curves.

The average standard score for the six basins in California was 1.279. These basins are characterized by thin, frequently unstable soils, resting on steep slopes in an active seismic region. The most distinctive feature which sets the California basins apart from the rest of the sample, however, is their Mediterranean climate. In all six basins, precipitation is strongly concentrated during the winter months from November to April and the summers are semiarid to arid. This seasonal contrast has been identified by Anderson *et al* (1959), Wilson (1972) and Jones *et al* (1972) as the major cause of the remarkably high sediment yields of most California streams. The very high sediment concentration during the intense winter storms and the low concentration at the onset of the dry season combine to produce typically steep rating curves with small intercepts. Thus the general rating functions (Equations 10-12) tend to overestimate the intercept and underestimate the slope for basins in California, apparently supporting Wilson's observation that the sediment transport regimes of basins in a Mediterranean climate are sufficiently different from the 'continental' regimes to require separate treatment.

The suggestion that the length of record used to determine the sample rating curves affects the accuracy of the general rating functions is based upon the progressive decrease in average standard score as record length increases (Table 4). There is, of course, no way of determining how representative an individual rating curve is; in a given instance, even a single year's record may fortuitously approximate the long-term average condition. However, it is reasonable to assume that the true sediment transport regime is in general more accurately reflected in longer records. From Table 4 it may be concluded that the general rating functions describe the long-term average sediment transport regime more accurately than is implied by the standard error of estimate for the entire sample.

Table 4   Effect of record length on accuracy of prediction of log k from Equation 12

Record length	Number of basins	Average standard score
1-4 years	13	1.018
4-7 years	18	0.762
7-10 years	9	0.635
greater than 10 years	10	0.495

7. Prediction of rating curves

DISCUSSION

The role of each variable in the general rating functions must be interpreted cautiously because of the empirical nature of the model and the simple set of variables used. Nonetheless, some general observations can be made.

From Table 3, 68% of the variance in log k is explained by mean annual runoff and relief, and mean annual runoff alone accounts for 47%. This is in accordance with the conclusions of Bauer and Tille (1967) who found a very high correlation between mean annual discharge and k for several East German streams. The reason for this is easily understood when it is recalled that k is the sediment discharge when Q = 1.0. As mean annual unit discharge decreases, then a discharge of 1.0 represents an event of increasing relative magnitude. Because sediment concentration is generally high for low frequency-large magnitude events, it follows that log k should vary inversely with mean annual unit discharge.

The addition of basin area to the equations resulted in only modest improvement in the levels of correlation, suggesting that the parameters which govern the sediment transported at a particular discharge are only weakly dependent upon the horizontal scale of the system. It may be inferred that basins which have similar vertical geometries and hydrologic regimes tend to have similar sediment transport characteristics as well, at least for the range of basin areas used in this study.

Channel slope and drainage density both explain relatively small amounts of the variance in log k and their inclusion in the general rating functions does not materially improve the accuracy of prediction. Channel slope is moderately well correlated with relief and this probably accounts for its small effect on the value of log k. Of interest, however, is the fact that slope is more significant when combined with H or s than it is in combination with $\bar{X}$. Channel slope, then, apparently acts as a surrogate for hypsometry in the equations which contain measures of range of relief and its inclusion probably contributes to the lack of significance of the hypsometric measures (skewness and kurtosis).

The slight effect of drainage density on the prediction of log k was somewhat surprising in view of the central role which drainage density has been shown to play in watershed dynamics. It may be that the diverse factors which control drainage density are not all related to the sediment transport process in the same way (or even in the same direction) and many of these factors may be self-cancelling or indeterminate when basins are compared on a continental scale. Thus, drainage density may not have a 'universal' significance in sediment transport studies. It is noteworthy however that with the partial exception of channel slope, drainage density is the only variable in the model which is internal to the system and adjustable and that what small effect drainage density has in the equation is in the opposite direction from the non-adjustable external factors.

The method used to measure drainage density further complicates the evaluation of its true or potential significance. The blue-line method was adopted by necessity and while its shortcomings were appreciated at the outset, it was felt that the method would indicate relative differences among basins. Nonetheless, the possibility that the small effect of drainage density on the equations reflects the inadequacy of the method must be considered.

The small contribution of channel slope and drainage density to the general sediment rating functions is ironic inasmuch as these variables are the most dependent upon the map scale for their derivation and thus placed the greatest restrictions upon the selection of basins for the study. By discarding channel slope and drainage density, the general rating functions can be described in terms of mean annual runoff, maximum or mean relief, and basin area with little sacrifice in the accuracy of prediction. From Table 3, the c and Z values for these new equations are as follows:

General sediment rating functions

$$G_s = c \, Z \, Q^{1.581 - 0.155 \log c - 0.155 \log Z}$$

where $c = 118.0; \quad Z = q^{-2.304} H^{-1.209} A^{-0.676}$ \hfill (13)

or where $c = 239.0; \quad Z = q^{-2.381} \bar{X}^{-1.511} A^{-0.677}$ \hfill (14)

Equations 13 and 14 explain 72-73% of the variance of log k and are much simpler to apply than the complete rating functions since they require only extensive basin properties which may be readily obtained from maps of any convenient scale. The probable errors associated with Equations 13 and 14 at 1.0 cfs are 285% and 274% respectively. At peak discharge for the sample basins, these errors reduced to 143% and 147%. Smaller errors might be expected when these equations are used to predict long-term sediment transport for basins outside the Mediterranean climate region.

The variables used to define Equations 10-14 were deliberately restricted to those which describe the morpho-climatic characteristics of watersheds in a general and universally-significant sense. The level of accuracy produced by these simple variables encourages the investigation of a larger set which more fully represents the physical nature of watershed processes. Particular attention might be paid to the quantitative representation of hillslope angles, to a more detailed description of the hydrologic regime, and to a better measure of the degree of channel development. The emphasis in the selection of additional variables, however, should remain upon those aspects of the watershed which fluvial systems have in common.

CONCLUSIONS

The relationship between the constant and exponent of suspended sediment rating curves permits the development of economical general suspended sediment rating functions based upon watershed characteristics. The level of accuracy achieved in this study using a simple variable set is comparable with the accuracy of other suspended load formulae which predict suspended bed material load only and encourages the investigation of a broader set of variables which more completely represents the nature of the watershed erosion-sediment transport system. In addition to their predictive value, general rating functions of the type proposed here would provide standards for the evaluation of sediment rating curves from a wide variety of environments.

ACKNOWLEDGEMENTS

The writer expresses his sincere appreciation to Dr. A. Jopling for his helpful criticism during the completion of this study and to the National Research Council of Canada for their financial assistance.

REFERENCES CITED

Anderson, H.W., Coleman, G.B. & Zinke, P.J., 1959, Summer slides and winter scour. *US Forest Service, Pacific Southwest Forest Range Experiment Station, Technical Paper* 36. 1-12

Bauer, L. & Tille, W., 1967, Regional differentiations of the suspended sediment transport in Thuringia and their relation to soil erosion. *International Association Scientific Hydrology Publications* 75, 367-377

Bogardi, J., 1961, Some aspects of the application of the theory of sediment transportation to engineering problems. *J. Geophysical Research,* 66, 3337-3346

Einstein, H.A., 1964, River sedimentation. In: Chow, V.T., ed., *Handbook of applied hydrology.* McGraw-Hill Book Co., New York

Evans, I.S., 1972, General geomorphometry, derivatives of altitude, and descriptive statistics. In: Chorley, R. J., ed., *Spatial analysis in geomorphology.* Methuen & Co. Ltd., London, 17-90

Jones, B.L., Hawley, N.L. & Crippen, J.R., 1972, Sediment transport in the western tributaries of the Sacramento River. *US Geological Survey Water Supply Paper* 1798-J

Leopold, L.B. & Maddock, T., 1953, The hydraulic geometry of stream channels and some physiographic implications. *US Geological Survey Professional Paper* 252

Shen, H.W., 1972, Total sediment load. In: Shen, H.W., ed., *River mechanics.* Colorado State University, Fort Collins, Colorado, 13.1 - 13.26

Tanner, W.F., 1959, Examples of departure from the Gaussian in geomorphic analysis. *American J. of Science,* 257, 458-460

Tanner, W.F., 1960, Numerical comparison of geomorphic samples. *Science,* 131, 1525-1526

Vanoni, V., ed., 1975, *Sedimentation engineering.* American Society Civil Engineers, New York, 745 pp

Wilson, L., 1972, Seasonal sediment yield patterns of US rivers. *Water Resources Research,* 8, 1470-1479

8 SUSPENDED SEDIMENT AND SOLUTE RESPONSE CHARACTERISTICS OF THE RIVER EXE, DEVON, ENGLAND

* D. E. Walling

ABSTRACT

River load investigations constitute an important aspect of the study of contemporary processes and denudation systems. This paper outlines some results obtained from a study of the River Exe and its tributaries and considers their wider implications. A network of thirteen primary measuring stations has been established in the study area and use has been made of apparatus for the continuous monitoring of specific conductance and suspended sediment concentrations and automatic pump sampling equipment.

A reconnaissance survey of the variation of total solute concentrations across the study area, carried out during a period of summer low flow, revealed a range of specific conductance levels in excess of 20-fold. The associated pattern has been related to control by rock type and land use. Mean annual solute loads have been evaluated for the thirteen measuring stations and the variations dis-cussed. Storm period solute response characteristics exhibit a typical dilution effect when total dissolved solids concentrations are considered, but a greatly increased com-plexity is demonstrated by the behaviour of individual ions.

Suspended sediment concentrations also exhibit marked spatial variability across the study area, and a 10-fold range of suspended sediment loads is apparent. Suspended sediment concentration/discharge relationships for the measuring stations do not demonstrate a clearly defined dependence of concentration on discharge, and hysteretic, seasonal and exhaustion effects on the relationships have been distinguished. In particular, sediment availability and the associated exhaustion effect exert an important control on sediment response during individual events, a sequence of hydrographs and during a longer period of several months.

The influence of aggregation and routing of sediment and solute response has been investigated using the data obtained from the monitoring network.

* University of Exeter, Devon, United Kingdom

The wider implications of these results include the relative importance of sediment and solute loads, spatial variation of process activity, the complex character of solute behaviour, the significance of response aggregation and routing, the importance of sediment availability in controlling sediment dynamics, and errors associated with the calculation of suspended sediment loads.

INTRODUCTION

Rates and modes of operation of contemporary processes have provided an important focus for recent geomorphological studies, both as a means of evaluating present-day denudation systems and of elucidating the development and modification of landforms. In particular, interest in fluvial processes has flourished over the past decade and the understanding of process mechanisms and interrelationships has advanced significantly. Within this general field, two major approaches can be distinguished. First, rates of lowering and modification of surface configuration, local removal rates and process dynamics have been investigated at individual sites or plots (eg. Campbell 1970; Imeson 1971; Smith & Wischmeir 1962; Soons & Rainer 1968). Secondly, erosion rates and processes have been inferred from measurements of the export of material out of a drainage basin by the stream. In this way, inventories of sediment and solute yield and denudation rates have been produced at the global level (Fournier 1960; Strakhov 1967), for individual countries (Slaymaker & McPherson 1973; Pulina 1972) and for local catchments (Smith & Newson 1974; Walling 1974). Combination of these two approaches should afford an improved understanding of the pattern and rates of operation of contemporary processes. Although attempted (Slaymaker 1972; Imeson 1974), increased refinement of measurement and interpretation is required before discrepancies can be meaningfully evaluated and the precise interaction of on-site erosion, river loads and delivery ratios defined.

Many problems exist in river load measurements and interpretation, including the field and laboratory techniques employed (Douglas 1971; Loughran 1971), the method used to calculate loads (Loughran 1976; Walling 1977), the representativeness of the data collected (Meade 1969), the assessment of non-denudational load components (Janda 1971) and appreciation of the temporal and spatial diversity represented by measurements at a single point. In several instances, use can be made of river load data collected by official organisations. Nevertheless, there remains a need for more detailed studies, in order to fully understand the underlying process mechanisms and to appreciate the potential and limitations of other available data. This was an underlying philosophy of a study of sediment and solute response characteristics within the basin of the River Exe, undertaken by the writer over the past five years.

Small catchments can provide invaluable information on sediment and solute processes, but the representativeness

of the results can often be questioned when extrapolation
to a wider area is sought. For this reason attention was
directed to an intermediate sized catchment (c 1000 km^2).
The catchment was selected to incorporate a considerable
degree of physical diversity and the measuring network was
structured to document the associated spatial variation in
response. Within the constraints inevitably imposed by
availability of labour and equipment, the monitoring programme was designed to provide meaningful information on
the short-term temporal characteristics of catchment
behaviour. This paper presents an overview of some of the
results obtained, and considers their wider implications
for the study of fluvial processes through river loads.

THE STUDY CATCHMENT

Description

The Exe basin (Figure 1) covers a drainage area of 1,462 km^2
and includes a diversity of rock types (Figure 2a) ranging
from the indurated slates and grits of the Devonian,
through the shales and sandstones of the Upper Carboniferous
to the marls, sandstones and breccias of the Permian and
Trias. The geological diversity is reflected by marked
contrasts in relief (Figure 1) which in turn condition
variations in hydrometeorological factors (Figure 2b).
Annual precipitation and runoff vary from over 1,800 mm and
1,500 mm respectively in the northern upland area of Exmoor
to less than 800 mm and 300 mm in the lowland area at the
head of the Exe estuary. Land use also varies across the
area with, for example, mixed farming characterising the
lowlands underlain by Permian and Triassic strata, permanent
pasture predominating on the Upper Carboniferous and the
appearance of some moorland on the higher areas of Exmoor.
From the viewpoint of the process geomorphologist, the
landscape must be considered largely relict. The major form
elements reflect development during the Tertiary, whilst
many of the details were added during the cold phases of
the Pleistocene. Contemporary processes are in many cases
merely reworking slope and valley floor deposits dating
from this latter period.

The absence of sizeable urban areas apart from Exeter
and Tiverton results in river loads that are not unduly
affected by effluent loadings. Loadings from non-point
sources associated with agricultural activity must exert
some influence on water quality, although the precise
extent is difficult to evaluate. The study area, therefore,
represents a relatively subdued process environment,
significantly influenced by agricultural activity but
nevertheless reflecting conditions characteristic of many
areas of Western Europe.

Instrumentation

Continuous river flow records are available for thirteen
measuring sites within the Exe catchment (Table 1). Eight
of these are river gauging stations maintained by the South
West Water Authority, whilst the remaining five are operated
specifically for this study. The resulting network (Figure 1)

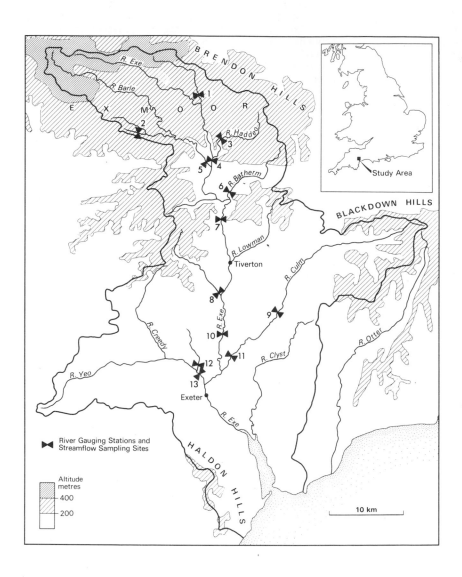

Figure 1   The basin of the River Exe showing the network of measuring stations

GEOLOGY

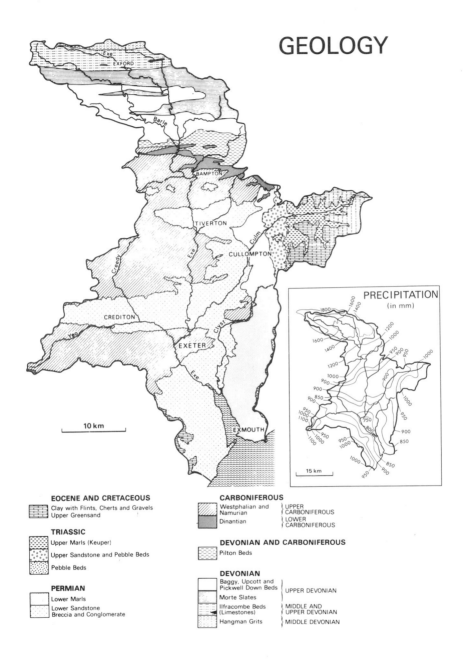

Figure 2   The geology and mean annual precipitation of the study area

provides a coverage of major tributaries and also affords
some opportunity to observe the behaviour of the catchment
in terms of aggregation of various subcatchment contributions.   Sampling at these gauging sites is carried out on
a regular weekly basis and more frequently during storm
events in order to document suspended sediment and solute
levels and their general pattern of variation through time.
Laboratory analysis of these samples includes determination
of suspended sediment concentrations, specific conductance
and the concentration of Ca, Mg, Na, K, NO_3 and Cl ions
using Atomic Absorption Spectrophotometry and an Autoanalyzer
system.   Sampling at the principal stations has been
supplemented by reconnaissance surveys at a larger number of
sites throughout the area for specific purposes.   A network
of five precipitation quality measuring stations is also
operated in the study area to provide information on bulk
precipitation inputs (Whitehead & Feth 1964).

Table 1 Measuring sites in the study area

	Site	Area (km^2)
1	R. Quarme at Enterwell	20.4
2	R. Barle tributary at Lyshwell	2.1
3	R. Haddeo at Hartford	50.0
4	R. Exe at Pixton	160.0
5	R. Barle at Brushford	128.0
6	R. Batherm at Bampton	64.5
7	R. Exe at Stoodleigh	422.0
8	R. Dart at Bickleigh	46.0
9	R. Culm at Woodmill	226.0
10	R. Exe at Thorverton	601.0
11	R. Culm at Rewe	273.0
12	R. Creedy tributary at Jackmoor	9.8
13	R. Creedy at Cowley	262.0

The manual sampling programme is inevitably insufficient
for detailed appraisal of temporal variations in sediment
and solute response, particularly during storm events when
continuous records or frequent sampling would be required.
Two complementary approaches have been adopted to provide
data for this purpose.   First, an attempt has been made to
continuously record sediment and solute concentrations.
Instrumentation for recording specific conductance has been
installed at twelve sites.   Specific conductance (SC)
closely reflects the ambient total dissolved solids (TDS)
concentration of a stream (Hem 1970) and establishment of
the associated relationships at each site has provided
clearly defined and consistent calibrations closely
approximating the form TDS (mg/l) = 0.65 SC (μs/cm 25°c).
Suspended sediment concentrations can fluctuate rapidly
during storm events and Guy (1965) has pointed to the

desirability of continuous monitoring apparatus to document
the fine-grained wash load. Technology is still lacking in
the provision of simple and reliable means of recording
sediment concentrations and, in the absence of a clear
alternative, use has been made of apparatus for continuous
monitoring of turbidity at nine of the gauging sites. As
in the case of conductivity, the sensing probes are mounted
directly in the river rather than in a pumped-cell system.
Some reports have pointed to the many problems involved in
the calibration of suspended sediment concentration against
turbidity (eg. Burz 1970). In the writer's experience,
however, very acceptable results can be obtained from streams
with a predominantly fine-grained suspended load providing a
field calibration is undertaken for each recording site.
The continuous record afforded by a turbidity sensor provides highly detailed information on the form of the sediment
graph and its precise time relation to the streamflow hydrograph.

The second approach to obtaining detailed temporal data
has involved the use of automatic pump sampling apparatus,
developed and constructed in the Geography Department of
the University of Exeter (Walling & Teed 1971). The current
design incorporates two collection units, one programmed to
collect regular samples at six hourly intervals, the other
operating at hourly intervals during storm events when the
river rises above an adjustable present level. Samplers of
this type are currently installed on the River Dart and
Jackmoor Brook (Sites 8 and 12, Table 1).

SOLUTE RESPONSE CHARACTERISTICS

Spatial variation:

Considerable spatial variation in the levels of solute concentration within the study area was apparent from an early
stage of the investigation, particularly between the low
concentrations exhibited by the streams draining from
Exmoor and the relatively high levels found in the southern
streams. Direct comparison of concentrations at different
sites is problematical because they vary according to the
relative discharge magnitude and the preceding flow conditions in a particular basin. Inter-site comparison was
thought to be most effective during the summer low flow
periods when concentrations are high but relatively stable.
Accordingly, samples were collected from over five hundred
locations within the Exe Basin during a twelve-day period
of stable low flows in the summer of 1971 (Figure 3a).
Because of the large number of samples involved, analysis was
restricted to specific conductance measurements which, as
noted previously, directly reflect the total solute content.
The sampling points were mainly on small tributaries and
the data have been used to construct the map of local
specific conductance levels presented in Figure 3b. On this
occasion, values varied from <40 μs/cm to >1,000 μs/cm, a
range of over 20-fold, and the pattern is closely related
to the underlying geology (Figure 2a). For example the
Devonian slates and grits exhibit the lowest values, the
Upper Carboniferous strata intermediate levels and the

175

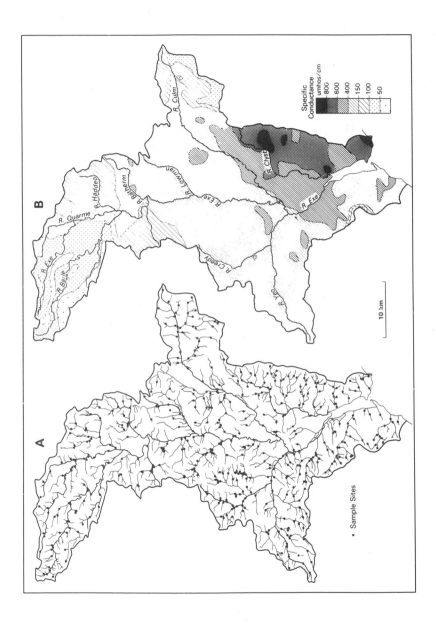

Figure 3   The network of sampling stations used in the 1971 reconnaissance survey
(A) and the resultant map of specific conductance levels for the survey
period (B)

Permian and Triassic rocks the highest levels. More
detailed examination of the values for individual small
catchments revealed that land use also exerted an influence.
Within the outcrop of the Upper Carboniferous, agricultural
catchments exhibited significantly higher conductivities
than forested basins and on the Devonian, moorland in turn
demonstrated lower values than woodland. In detail the
pattern reflects the complex interaction of many environmental controls (Walling & Webb 1975).

Table 2   Average annual total solute loads
           calculated for the 13 measuring stations

Site	Annual load	
	(tonnes)	(tonnes/km^2)
1	1872	91.8
2	62	29.4
3	3209	64.2
4	12050	75.3
5	6785	53.0
6	5908	91.6
7	29013	68.8
8	2795	60.7
9	23222	102.8
10	50905	84.7
11	26112	95.6
12	627	64.0
13	16050	61.3

The mean annual solute loads have been estimated for
the thirteen gauging stations (Table 2) using the streamflow duration curves and total dissolved solids concentration/
streamflow relationships (eg. Figure 4). The values exhibit
some variation but this is not as marked or as closely
related to rock outcrops as that shown by concentration
levels. Variations in runoff across the area reduce the contrasts associated with concentration, because catchments with
low TDS values are generally associated with high values of
annual runoff. Thus the Lyshwell catchment on Exmoor
(Site 2, Figure 1) has the lowest solute load - 29t/km^2 and
the neighbouring Quarme (Site 1, Figure 1) also draining an
area of Exmoor, exhibits the third highest load - 91.8t/km^2

Temporal Variation:

Investigations of temporal variations in solute response are
required to provide an insight into the process mechanisms
operating. The solute concentration/streamflow relationship is often used to provide a general representation of
temporal behaviour and the plots for a representative

177

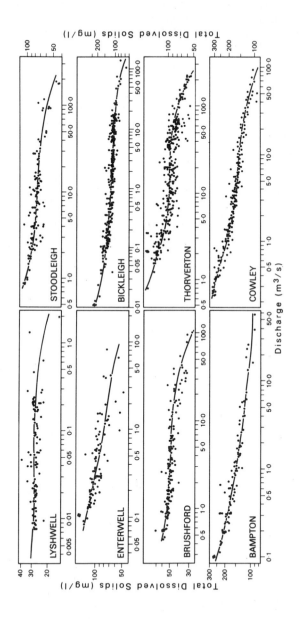

<u>Figure 4</u> Total solute concentration/discharge relationships
for a number of measuring stations

selection of gauging sites are presented in Figure 4. All
demonstrate the commonly occurring inverse relationship of
concentration to discharge, associated with the dilution of
relatively high baseflow solute levels by the lower concentrations of runoff with shorter residence times. The
extent of variation in concentration between high and low
discharges varies considerably between sites. The Lyshwell
station, on a moorland tributary of the River Barle, shows
little variation, whereas the data for the River Creedy at
Cowley exhibit a 3-fold range. The general solute levels
associated with the individual sites reflect solute
availability and controls similar to those associated with
the spatial variation of baseflow concentrations depicted
in Figure 3b. The slope of the rating plot is conditioned
by the range of flows experienced within a catchment and
the contrast between runoff with long and short residence
times, which in turn relates to the buffering capacity of
the soil and bedrock and solute availability.

Examples of the dilution of solute concentrations, as
reflected by specific conductance measurements, during
individual storm events, are presented in Figure 5. The
four catchments exemplify the range of response encountered
in the study area. At the Lyshwell station, the storm
hydrograph generates little change in specific conductance
and the lack of contrast between stormflow and baseflow
can be attributed to the low solute availability within
this catchment. Atmospheric contributions dominate the
solute supply, since preliminary calculations of chemical
denudation rates for this catchment, based on subtraction
of atmospheric inputs from gross solute loads, provide
zero values. Stream solute levels therefore reflect the
quality of incoming precipitation and dry fallout rather
than solute pickup within the basin and there will be
little contrast between stormflow and baseflow. Conversely,
terrestrial sources dominate the solute response of the
River Creedy tributary and a marked dilution is apparent
during storm events. The Rivers Dart and Quarme provide
examples where the contrast between baseflow and stormflow
is intermediate between the two previous catchments.
Detailed study of specific conductance records has also
revealed variations in the relative timing of the streamflow
peak and trough of total solute concentration. In Figure 5,
the two are seen to correspond closely but records from
the River Creedy reveal a variable lag between the discharge
peak and the solute trough, with the latter occurring
0-14.5 hours after the former (Walling & Foster 1975). This
variation in timing has been related to antecedent moisture
conditions within the catchment.

The behaviour exhibited by TDS levels in the con-
centration/discharge plots (Figure 4) and the response of
individual storm events (Figure 5) lends itself to interpretation in terms of a simple dilution model. Simple flow
component mixing models have been used successfully to
explain temporal variation in specific conductance levels
(Walling 1974). However, care must be exercised in
developing such models to provide a meaningful representation
of detailed solute processes, since TDS and conductivity

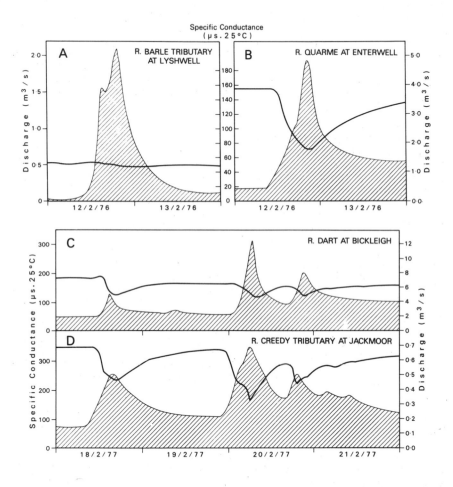

<u>Figure 5</u>   The trend of specific conductance levels
during individual storm hydrographs at
four of the measuring stations

values provide a lumped or aggregate measure. For example,
the specific conductance trace illustrated for three
successive storm events on the River Dart in Figure 6b
demonstrated limited variation and could be interpreted
as reflecting a buffering effect and a general lack of contrast between stormflow and baseflow. Inspection of Figure
6c reveals the complexity of the behaviour of several
individual ions during this period. It is not the purpose
of the present discussion to analyse their behaviour in
detail but emphasis of the complexity and comment on the
need to interpret the response of specific ions individually
in terms of sources, mobility and runoff dynamics is
appropriate.

In this example Mg levels alone demonstrate a simple
response consistent with a simple dilution model. Potassium
concentrations contrast markedly with the behaviour of all
other ions in showing an increase during storm events,
superimposed on a progressive decrease in the background
level through the six day period. This pattern is somewhat
difficult to explain, although similar findings have been
reported elsewhere, for example by Steele (1968) for
Pescadero Creek, California, and by Spraggs (1976) for the
West Walk catchment in Hampshire, UK. Considerations of a
surface source, interplay of residence times and ion
exchange and adsorbtion and interaction with the suspended
sediment load must be introduced.

Chloride can be viewed as primarily atmospheric in
origin within this catchment, and the close proximity of
the sea (c 35 km) accords an obvious oceanic source. Its
behaviour is generally viewed as conservative and will
reflect the physical process of input, storage, moisture
evaporation and mixing. The two 'flushes' of increased concentration preceding the storm-period dilution can be
attributed to removal of accumulated deposits from vegetation
and the surface and upper levels of the soil. It is
attractive to attribute the gradual decrease in Cl concentrations following the two major storm hydrographs to the
progressive appearance of water with shorter residence times
by the process of translatory flow (eg. Hewlett & Hibbert
1967). Cl/Na ratios in this catchment approximate 1.8,
the level ascribed to an oceanic source, although the two
ions demonstrate certain contrasts in response attributable
to the less conservative behaviour of Na. Calcium can be
considered to be essentially terrestrial in origin, and
the pronounced flushing effect preceding the second major
streamflow rise provides a very distinctive feature of its
behaviour.

Stream solute levels reflect not only geochemical processes but also biotic control, since many solute species
are involved in nutrient cycling. Nitrate provides one such
good example and its availability is further influenced by
the rate of mineralisation of organic nitrogen. In Figure
6c the NO_3 trace contrasts strongly with that of the other
ions in that it demonstrates a marked increase following
the storm hydrograph. The concentration peak occurs thirty-
siz hours after the streamflow peak and is apparently related
to sub-surface flow through the upper soil horizons, for

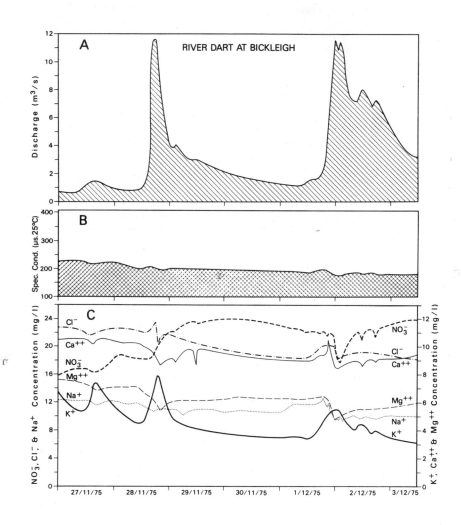

Figure 6   Variation of streamflow (A), specific
conductance (B) and the concentration
of individual ions (C) during a series
of storm events on the River Dart

the timing is reminiscent of the throughflow peaks described by other workers (Anderson & Burt 1977; Hewlett & Nutter 1970; Troake & Walling 1973). Dilution is clearly evident during the subsequent streamflow peak and the immediate dilution associated with direct runoff can be contrasted with the delayed increase related to subsurface throughflow.

The detailed solute behaviour exhibited by this and other catchments in the study area clearly demonstrates the limitation of a TDS or specific conductance record in revealing process mechanisms. These will be masked by the aggregation of the contrasting time-variant response of individual ionic species.

SUSPENDED SEDIMENT RESPONSE CHARACTERISTICS

Spatial Variation:

Because suspended sediment concentrations only increase to significant levels during storm runoff events and then fluctuate rapidly, it is impossible to extend the reconnaissance survey approach used for solute levels to document spatial variation in sediment concentrations. Nevertheless, samples collected at the primary measuring stations (Figure 1) have revealed considerable variations in storm-period sediment concentrations across the study area. In the moorland Lyshwell catchment (Site 2, Figure 1) levels rarely exceed 50 mg/l and at Brushford on the Lower Barle (Site 5) values as high as 200 mg/l are rare. In contrast concentrations in excess of 3,000 mg/l have been recorded on several occasions on the River Dart (Site 8) and the Jackmoor Brook (Site 12). Because of the problems involved in obtaining meaningful estimates of sediment load, discussed later, values are currently presented only for the Rivers Dart and Creedy and the Exe at Thorverton for limited periods (Table 3). These provide a first approximation of the suspended sediment loads characteristic of the study area. The Dart load (90.8 t/km^2) must approach the highest in the area, whereas the northern tributaries must possess lower values than those for the Exe at Thorverton, probably less than 10 t/km^2. A 10-fold range of sediment loads therefore characterises the area.

Table 3   Suspended sediment loads calculated for three measuring stations using the continuous turbidity records

River	Period	Annual load (tonnes)	(tonnes/km^2)
Creedy	Oct 72 - Sept 74	27990	53.4
Dart	Jan 75 - Dec 75	4179	90.8
Exe	Oct 74 1 Sept 75	14442	24.0

Temporal Variation:

In many investigations the suspended sediment concentration/discharge plot has been used to characterise the response of a basin and to infer process mechanisms. One characteristic of the streams sampled in this study is the lack of

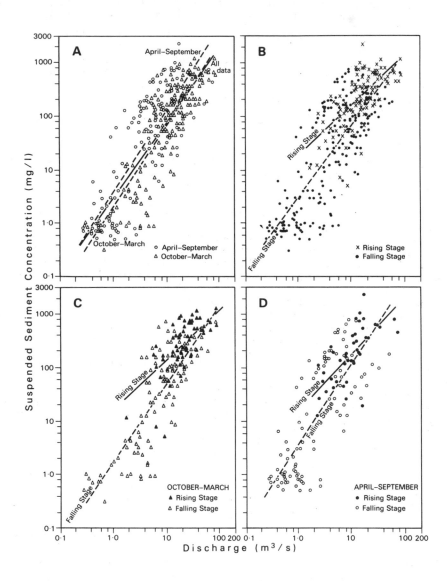

Figure 7   Suspended sediment concentration/discharge
relationships for the River Creedy at
Cowley, distinguished according to season
(A), stage (B) and stage and season (C & D)

well defined concentration/discharge relationships. As an example, the rating plot for the River Creedy is presented in Figure 7a. This has been based on regular weekly sampling supplemented by additional measurements during storm events. The least squares line fitted to the logarithmic plot accounts for only 69% of the variance of the concentration data (Table 4) and a considerable band of scatter is apparent. Previous workers (eg. Hall 1967; Nilsson 1971) have attempted to improve the relationship by subdividing the data set according to season and to the relationship of the sampling time to the streamflow hydrograph in terms of rising or falling stage. In Figure 7 the plots for the sub-divided data sets are presented and the form and goodness of fit of the associated regression lines are listed in Table 4. The River Creedy, like other streams in the area, is characterised by a tendency for sediment concentrations for a given discharge to be higher during the summer (April-September) period than during the winter (October-March) and also to be higher on the rising limb of a hydrograph than the falling stages. However, use of these individual relationships provides little improvement in explanation over the relationship for the total data set. No clear relation between concentration and discharge, indicative of a simple transport function, is apparent for this or other rivers in the area.

Table 4   Equation form and goodness of fit for the various sediment concentration/discharge relationships developed for the River Creedy

Rating relationship	a	b	r	n
All data	3.087	1.408	0.83	309
Summer	3.749	1.501	0.83	138
Winter	1.950	1.498	0.84	171
Rising stage	16.27	0.944	0.70	103
Falling stage	2.742	1.348	0.80	206
Summer rising	13.52	1.067	0.70	42
Summer falling	3.451	1.467	0.78	96
Winter rising	15.88	0.930	0.70	61
Winter falling	1.675	1.457	0.83	110

where: a, b = constant and coefficient in equation
$$conc = aQ^b$$
r = correlation coefficient
n = number of observations

Inspection of the data obtained from the continuous turbidity recorders installed at a number of gauging stations provides an insight into the causes of the lack of a well-defined rating relationship. In Figure 8 the suspended sediment response through two storm events on the Barle, Creedy and Dart is portrayed. In all cases the concentration/discharge relationship exhibits hysteresis, with concentrations of the rising limb exceeding those for a similar discharge on the recession limb of the hydrograph. This pattern is further complicated in the case of the Dart and Creedy by contrasts in the timing of the discharge and concentration peaks. The peak sediment concentration precedes the

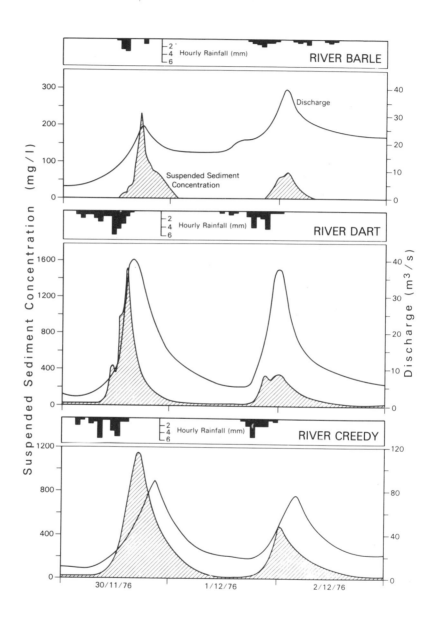

Figure 8   Variation of suspended sediment
concentration at three of the
measuring stations during individual
storm events

maximum streamflow by several hours and concentrations
have fallen to about 50% of the peak value by this time.
Lack of coincidence in timing could be attributed to the
routing of runoff from various contributing areas past the
measuring station, with an early sediment peak occurring
if the lower portions of the catchment provided the major
sediment source. In these catchments, however, it is
thought more likely to reflect the exhaustion of the supply
of available sediment as the storm event proceeds, a
feature reported by other workers in the context of soil
erosion studies (Ellison 1945; Emmett 1970). Associated
with this feature and also demonstrated in Figure 8 is the
tendency for the concentration levels exhibited by a second
storm of similar magnitude, occurring in close succession,
to be considerably lower than those of the first. In
certain instances this might be attributable to contrasts
in rainfall intensity but this factor is not immediately
evident in the examples presented and a progressive
exhaustion of the sediment supply would again seem to be
operative. Care must be exercised in interpreting some
apparent instances of exhaustion of sediment supply through
a series of successive runoff peaks. Where a subsequent
hydrograph is superimposed on a considerably increased
baseflow, the sediment produced by the storm runoff component
will be diluted by the high baseflow to give an apparent
exhaustion effect. This feature is significant in the
examples presented in Figure 8 but is not the major cause.
The suspended sediment dynamics of the study area are
dominated by supply conditions which can exibit a rapid
exhaustion effect during a single hydrograph and through a
series of events.

The existence of an exhaustion effect during a
succession of storm hydrographs is further clearly demonstrated in Figure 9a for the River Dart. In this sequence
there would appear to be some recovery of availability
between the second and third peaks but by the fifth hydrograph the increase in concentration is almost negligible.
A general exhaustion effect can also be apparent over a
longer period of time, for example as a season progresses,
and an attempt to exemplify this is presented in Figure 9b.
This considers a series of forty-two clearly defined storm
hydrographs on the River Dart during the period September
1976 to February 1977 following the drought conditions of
the previous summer. The peak concentration for each
event (C) has been plotted against the increase in discharge or hydrograph rise associated with that event (Qr).
The latter variable provides something of a surrogate for
erosive energy and the relationship has previously been
found useful in explaining storm period sediment yields ·
(Walling 1974). The values of the four storms occuring
between 11th and 22nd September 1976 could be viewed as
defining an approximate relationship of the form C ∝ Qr
with successive storms during the period 22 September to
3 October 1976 plotting progressively further to the right
and indicating a decrease in sediment availability and a
parallel shift of the relationship. Some recovery of supply
is apparent between 3 October and 11 October 1976, but

8. River Exe, Devon, England

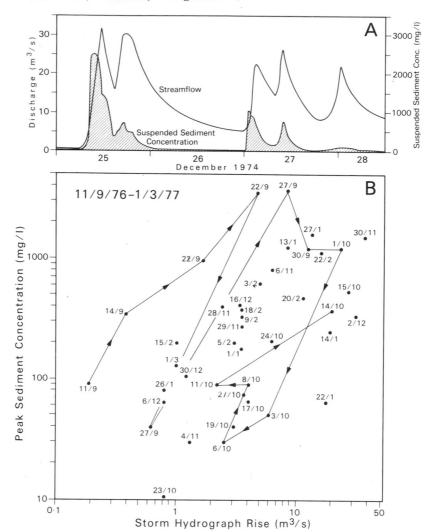

Figure 9   The variation of suspended sediment
concentration during a succession of
five storm hydrographs on the River
Dart (A) and the relationship between
peak sediment concentration and storm
hydrograph rise (B) for individual
events occurring between 11.9.76 and
1.3.77 on the same river. The points
for the first thirteen events have
been linked to demonstrate the trend
of the relationship during that
period

8. River Exe, Devon, England

availability has decreased again by the storm of 14 October 1976. A similar pattern of exhaustion and recovery can be traced through the remaining events.

In a first attempt to quantify the exhaustion effect apparent in Figure 9b, multiple regression has been employed to derive a relationship between peak sediment concentration (C) the hydrograph rise (Qr) and a sediment availability index (Ai) based on antecedent discharge conditions. This index is analogous to an API index and reflects both the magnitude and the proximity in time of previous events. It is calculated on a daily basis by adding the increase in discharge associated with storms and utilising an exponential decay function to reduce the effect of that and previous events as time proceeds: ie.

$$Ai = (Ai_{d-1}K) + \Sigma Qr$$

Where: Ai = sediment availability index at end of day

Ai_{d-1} = sediment availability index on previous day

K = constant decay

Qr = storm discharge increment or hydrograph rise on the given day

The value of 0.9 has been assigned to K, although more work is required to ascertain the optimum value.

Table 5   Multiple regression relationship derived between peak sediment concentration for individual events and the associated values of hydrograph rise and the sediment availability index

$Log\ C$ = $2.4454 + 0.8825\ Log\ Qr - 0.4905\ Log\ Ai$

R = 0.80

n = 42

Where C = peak suspended sediment concentration for an individual event (mg/l)

Qr = hydrograph rise for event (m³/s)

Ai = availability index

The resulting equation is presented in Table 5. A stepwise routine was used to derive the relationship and Qr entered the regression first, accounting for 30% of the variance of C, whilst Ai' entered second bringing the total explanation to 64%. The importance of sediment availability and an exhaustion effect through a series of events is therefore further demonstrated by the importance of the Ai variable in this relationship. Availability seems equal in importance to discharge conditions in controlling sediment concentrations in this area and scope exists for refinement of the Ai index based on a continuous accounting procedure.

8. River Exe, Devon, England

THE AGGREGATION AND ROUTING OF
SEDIMENT AND SOLUTE RESPONSE

When the spatial and temporal characteristics of sediment
and solute delivery of a river are investigated with a
view to interpreting the process dynamics of the catchment
upstream, the influence of aggregation of the response of
the several subcatchments and of downstream routing on the
behaviour of sediment and solute concentrations, particularly
during storm events, must be considered. The availability
of data from a series of measuring stations within the same
river network affords an opportunity to analyse this
phenomenon further.

Data collected from the River Exe at Thorverton
(Site 10, Figure 1) can be more meaningfully interpreted
if the response is considered as the resultant of the output
from various subcatchments rather than the behaviour of
a lumped catchment unit. For example, the sediment concentration associated with individual floods can vary significantly according to whether the flood discharge was
generated uniformly across the basin or more locally in
either the northern headwaters or the southern area. Flood
discharges generated in the southern tributaries exhibit
relatively high sediment concentrations compared to flows
originating on Exmoor. Similarly, marked variations in the
timing of the peak of sediment concentration relative to
the streamflow can result. On certain occasions the peak
sediment concentration during a flood event on the Exe
precedes the discharge peak by eight hours and can be
related to the sediment load from the River Dart. In other
events, where the northern area dominates as a source of
floodwater, the sediment peak may only precede the flood
peak by two hours or less, although a minor peak in sediment
concentration occurring about six hours previously and
related to the contribution from the Dart can generally be
distinguished. Similar features are evident in the solute
response recorded at Thorverton.

Equally significant are the effects of downstream
routing and in this context analysis of data collected
from the River Culm has been undertaken. On this river
there are two measuring stations 13 km apart (Figure 1).
Only one relatively small tributary, the River Weaver, joins
in the intervening reach which flows through a well
developed floodplain with many meanders (Figure 10). Figure
10 presents the streamflow hydrograph and sediment and
solute response for two successive flood events measured
at the upstream (Woodmill) and downstream (Rewe) stations.
The graph of sediment concentration exhibits considerable
smoothing as it proceeds downstream. Individual peaks (A)
evident at Woodmill are marked by slight flattenings at
Rewe, although the input from the River Weaver shows up
clearly (B). The 13 km reach between the two stations is
an area of active meander development and bank erosion rates
of 0.2 m/yr have been reported (Hooke 1977). Initially it
was thought that this might be an important sediment source
for this river, but calculations indicate that sediment
loads tend to decrease over the reach. In the two events

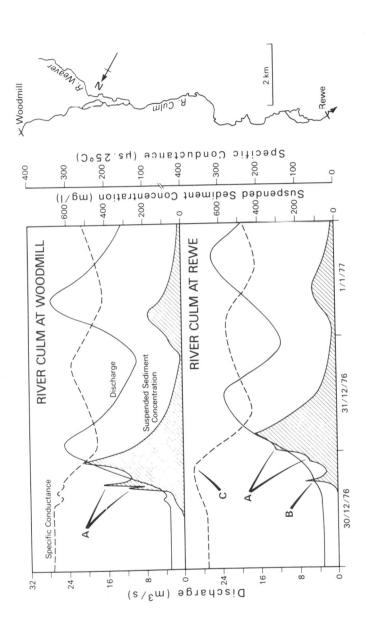

Figure 10 Comparison of the streamflow, suspended sediment and specific conductance response at Woodmill (upstream) and Rewe (downstream) on the River Culm

illustrated, downstream loads were only 74% and 91% respectively of the upstream loads and any increases in loads provided by bank erosion and the River Weaver are more than countered by the losses of sediment in this reach. The sediment delivery ratio is an important parameter for the geomorphologist attempting to interpret stream loads in terms of erosion rates and this example demonstrates the very significant reduction in delivery rates that can occur along a relatively short channel reach.

Changes in the form of the specific conductance trace also occur between the two measuring sites and, in particular, the small bump which preceeds the dilution trough at the downstream site can be related to floodlfow contributions from the River Weaver which exhibits relatively high conductivities. The influence of overbank spillage and floodplain storage on the form of the flood hydrograph also influences the relative timing of the water, sediment concentration and specific conductance peaks. Whereas the sediment peak precedes the discharge maxima by approximately four hours, and the conductivity trough is almost coincident with the water peak at the upstream station, at the downstream site the sediment peak and the specific conductance trough precede the flood peak by eight hours and three hours respectively. This occurrence is the reverse of that found by Heidel (1956) for sediment, and by Glover & Johnson (1974) for solute response. They describe a progressive lag of sediment and solute response downstream consequent upon the process of flood wave generation.

SOME IMPLICATIONS

The study from which some of the findings have been outlined above is to a large extent an empirical investigation of fluvial processes in a local area. Nevertheless, some wider implications are apparent and a number of these can be introduced.

1) The relative importance of solute loads and chemical denudation

A detailed appraisal of the relative magnitude of suspended sediment and solute loads in the study area is not currently feasible due to the lack of long-term sediment data. However, the data presented in Table 2 and Table 3 suggests that the two are approximately equal for the River Creedy and that sediment exceeds solute load for the Dart, whilst for the Exe at Thorverton, solute load outweights sediment load by 3-4 times. The solute/sediment load ratio probably increases to >5 on some of the Exmoor tributaries. Over the Exe basin as a whole, solute loads appear to be considerably more important than suspended sediment yields. An attempt has been made to convert the gross solute loads to provisional estimates of chemical denudation, by subtracting the atmospheric contribution. Although this approach has many shortcomings (Janda 1971), the results, converted to volume removal rates by assuming a specific gravity factor of 2.5, are presented in Table 6. If the

available sediment data are similarly adjusted to take into
account the organic fraction (a mean value of 15% has been
assumed) and the same specific gravity, the results may be
compared (Table 6). Chemical denudation exceeds suspended
sediment loss by a factor of nearly 3 for the basin of the
Exe above Thorverton. The different delivery ratio
characteristics of sediment and solute loads should, however, be considered when making further inferences concerning upstream erosion and denudation rates.

Table 6   Comparison of chemical denudation rates
           and sediment loss in the study area

Measuring station	Chemical denudation	Sediment loss
1	23.1	-
2	0.0	-
3	14.0	-
4	17.4	-
5	7.4	-
6	26.1	-
7	15.4	-
8	15.4	30.9
9	31.5	-
10	22.3	8.2
11	29.0	-
12	18.0	-
13	16.1	18.1

Traditionally the importance of chemical denudation has
been emphasised for limestone areas, but it must be recognised that they may be very significant in areas of other
lithology.

2) Spatial variability in catchment response

Spatial variability in the operation of catchment processes
can be viewed at a variety of scales from that of the small
drainage basin, where the partial area model of runoff
production has important implications (Engman 1974), to that
of the large region characterized by physiographic diversity.
This study has highlighted the variation that can occur
within an intermediate sized catchment and serves to
emphasise the need for caution in interpreting the sediment
and solute yields as lumped values characteristic of the
watershed upstream.

3) The complex character of solute response

Measures of total dissolved solids concentration and
specific conductance provide an effective means of docu-
menting variation in solute yield from a drainage basin.
However, their totalising nature effectively masks much
of the complexity of solute behaviour. If solute processes
are to be inferred from river loadings, intensive sampling
and investigation of the response of individual ions in

terms of sources, sinks and runoff dynamics will be required.

4) The significance of response aggregation and routing

A detailed appreciation of this aspect of sediment and solute behaviour will also be required if catchment-wide processes are to be inferred from river transport measurements. Furthermore, downstream transmission losses may be highly significant when evaluating suspended sediment loads and delivery ratios.

5) The importance of sediment availability

Previous work on interpreting and modelling the process of sediment entrainment and transport has relied heavily on plot and laboratory studies and has focused on defining the process mechanisms and relating their efficiency to precipitation character and runoff hydraulics. Erodibility is in many cases treated as an essentially static factor. Investigations of river loads carried out in this local area have demonstrated that attention must be given to variations in sediment availability through time and the exhaustion effect, if meaningful models are to be developed. A detailed availability accounting procedure, based on catchment moisture status and antecedent precipitation and discharge may be required.

6) The calculation of suspended sediment loads

In areas where exhaustion effects constitute an important characteristic of suspended sediment response, sediment concentration/discharge relationships will generally be poorly defined and will exhibit considerable scatter. Use of these relationships to calculate sediment loads from the streamflow record, using the sediment rating curve approach, could give rise to significant errors. Further-more, it is an inherent assumption of the rating curve approach that the water and sediment peaks are synchronous and that there is a simple relationship between the form of the runoff and sediment concentration hydrographs. In many cases the majority of the sediment may move as a slug of short duration compared to the streamflow event.

To eludicate the potential errors associated with the use of rating curves, a comparison has been made between values of sediment load for the River Creedy, calculated using the continuous turbidity record and those calculated using concentration/discharge relationships. Both a single relationship and ratings subdivided according to season and stage tendency (Figure 7) have been used. Hourly discharge data have been utilised in both cases. Results are presented in Table 7, and in all cases use of a rating curve considerably overestimates the sediment load. Errors of +50% may be involved, and such potential error terms should be borne in mind when sediment load data involving different computation procedures are used in further statistical analysis and when sediment and solute loads are compared. The comparisons presented in this paper are based on sediment loads calculated using the continuous turbidity records, different results would have resulted if the rating curve-derived estimates had been employed.

Table 7   Comparison of suspended sediment loads
         for the River Creedy calculated using the
         continuous turbidity record and sediment
         concentration/discharge relationships

Method of cal- culating load	Load Oct 72-Sept 74	Error involved in use of rating curve
Continuous tur- bidity record	27990 tonnes	-
Single rating	43078 tonnes	+ 53%
Seasonal ratings	41672 tonnes	+ 49%
Stage ratings	34607 tonnes	+ 24%
Combined stage and seasonal ratings	33996 tonnes	+ 21%

ACKNOWLEDGEMENT

The writer gratefully acknowledges the financial assistance
provided for work in the Exe Basin by the Natural
Environment Research Council. The map presented in Figure 3
is based on a survey carried out with Mr. B. W. Webb, and
his cooperation with this and other aspects of the work
described is acknowledged with thanks.

Thanks are also due to Mr. R. Carter, Mr. J. Bligh,
and The South West Water Authority, who have provided
valuable assistance with data collection and laboratory
analysis.

REFERENCES CITED

Anderson, M.G. & Burt, T.P., 1977, Automatic monitoring of
    soil moisture conditions in a hillslope spur and
    hollow. *J. Hydrology*, 33, 27-36

Burz, J., 1970, Experience with photometric turbidity
    measurements. *Int. Ass. Hydrol. Sci. Publication* 99,
    519-530

Campbell, I.A., 1970, Erosion rates in the Steveville bad-
    lands, Alberta. *Canadian Geographer*, 14, 202-216

Douglas, I., 1971, Comments on the determination of fluvial
    sediment discharge. *Australian Geographical Studies*,
    9, 172-176

Ellison, W.D., 1945, Some effects of raindrops and surface
    flow on soil erosion and infiltration. *Trans.
    American Geophysical Union*, 26, 415-429

Emmett, W.W., 1970, The hydraulics of overland flow on
    hillslopes. *US Geol. Survey Prof. Paper* 662A

Engman, W.T., 1974, Partial area hydrology and its appli-
    cation to water resources. *Water Resources Bulletin*
    10, 512-521

8.   River Exe, Devon, England

Fournier, F., 1960, *Climat et erosion: la relation entre l'erosion du sol par l'eau et les precipitations atmospheriques.* Presses Universitaires de France, Paris, 201 pp

Glover, B.J. & Johnson, P., 1974, Variations in the natural chemical concentration of river water during flood flows and the lag effect. *J. Hydrology,* 22, 303-316

Guy, H.P., 1965, Fluvial sediment measurements based on transport principles and network requirements. *Int. Ass. Hydrol. Sci. Publication* 67, 395-409

Hall, D.G., 1967, The pattern of sediment movement in the River Tyne. *Int. Ass. Hydrol. Sci. Publication* 75, 117-140

Heidel, S.G., 1956, The progressive lag of sediment con- centration with flood waves. *Trans. American Geophysical Union,* 37, 56-66

Hem, J.D., 1970, Study and interpretation of the chemical characteristics of natural water. *US Geol. Survey Water Supply Paper* 1473, 363pp ˅

Hewlett, J.D. & Hibbert, A.R., 1967, Factors affecting the response of small watersheds to precipitation in humid areas. In: Sopper, W.E. & Lull, H.W. (eds) *International Symposium Forest Hydrology,* Pergamon Press, 275-290

Hewlett, J.D. & Nutter, W.L., 1970, The varying source area of streamflow from upland basins. In: *Proceedings of a Symposium on the Interdisciplinary Aspects of Watershed Management, ASCE,* 65-83

Hooke, J., 1977, Personal communication

Imeson, A.C., 1971, Heather burning and soil erosion on the North Yorkshire Moors. *J. Applied Ecology,* 8, 537-542

Imeson, A.C., 1974, The origin of sediment in a moorland catchment with particular reference to the role of vegetation. *Institute of British Geographers Special Publication* 6, 59-72

Janda, R.J., 1971, An evaluation of procedures used in computing chemical denudation rates. *Geological Society America Bull.,* 82, 67-80

Loughran, R.J., 1971, Some observations on the determination of fluvial sediment discharge. *Australian Geographical Studies,* 9, 54-60

Loughran, R.J., 1976, The calculation of suspended sediment transport from concentration v discharge curves: Chandler River, NSW. *Catena,* 45-61

Meade, R.H., 1969, Errors in using modern stream-load data to estimate natural rates of denudation. *Geological Society America Bull.,* 80, 1265-1274

Nilsson, B., 1971, Sediment transport i svenska vattendrag Ett lHD - projekt. Del. 1: Metodik. *UNGI Rapport,* 4, Uppsala

Pulina, M., 1972, A comment on present-day chemical denudation in Poland. *Geographica Polonica,* 23,45-62

Slaymaker, H.O., 1972, Patterns of present sub-aerial erosion and landforms in mid-Wales. *Trans. Institute of British Geographers,* 55, 47-68

Slaymaker, H.O. & McPherson, H.J., 1973, Effects of land use on sediment production. In: *Fluvial Processes and Sedimentation,* Canada National Research Council, 159-183

Smith, D.D. & Wischmeier, W.H., 1962, Rainfall erosion. *Advances in Agronomy,* 14, 109-148

Smith, D.I. & Newson, M.D., 1974, The dynamics of solutional and mechanical erosion in limestone catchments on the Mendip Hills, Somerset. *Institute of British Geographers Special Publication* 6, 155-168

Soons, J.M. & Rainer, J.N., 1968, Micro-climate and erosion processes in the Southern Alps, New Zealand. *Geografiska Annaler* 50A, 1-15

Spraggs, G., 1976, Solute variations in a local catchment. *The Southern Hampshire Geographer,* 8, 1-14

Steele, T. D., 1968, *Seasonal variation in chemical quality of surface water in the Pescadero Creek watershed, San Mateo County, California.* PhD dissertation, Stanford University

Strakhov, N.M., 1967, *Principles of lithogenesis, 1.* Oliver & Boyd, New York, 245pp

Troake, R.P. & Walling, D.E., 1973, The hydrology of the Slapton Wood Stream. A preliminary report. *Field Studies,* 3, 714-740

Walling, D.E., 1974, Suspended sediment and solute yields from a small catchment prior to urbanization. *Institute of British Geographers Special Publication* 6, 169-192

Walling, D.E., 1977, Limitations of the rating curve technique for estimating suspended sediment loads, with particular reference to British Rivers. In: *Proc. International Symposium on Erosion and Solid Matter Transport in Inland Waters, Paris, Int. Ass. Hydrol. Sci. Publication*

Walling, D.E. & Foster, I.D.L., 1975, Variations in the natural chemical concentration of river water during flood flows and the lag effect: some further comments. *J. Hydrology,* 26, 237-244

Walling, D.E. & Webb, B.W., 1975, Spatial variation of river water quality: a survey of the River Exe. *Trans. Institute of British Geographers,*65,155-171

Whitehead, H.C. & Feth, J.H., 1964, Chemical composition of rain, dry fallout and bulk precipitation at Menlo Park, Calif., 1957-59. *J. Geophysical Research,* 69, 3319-3333

9 CHEMICAL AND MINERALOGICAL INDICES OF SEDIMENT TRANSFORMATION DURING FLUVIAL TRANSPORT

* Terry J. Logan

ABSTRACT

The chemistry and mineralogy of suspended sediments in fluvial transport are discussed as indices of sediment source area and as indicators of surficial and streambank erosion. Changes in sediment chemistry and mineralogy during fluvial transport are discussed, and the role of sediments in transporting pollutants and regulating the soluble forms of pollutants is described. The role of sediment chemistry and mineralogy in clay coagulation and elemental adsorption/desorption in the fluvial environment is highlighted.

INTRODUCTION

Early studies on fluvial transport of sediment were concerned with the hydraulics of sediment movement. Total sediment load, bed load, etc. were studied to determine the dynamics of sediment movement in streams. Particle size analysis of stream sediments was determined to characterize sediment deposition patterns, stream stability, reservoir siltation and other fluvial processes. The growing concern of the 1960's for cultural eutrophication of our lakes and biological degradation of our rivers and streams has led to renewed examination of sediment transport phenomena, but with a new objective: to understand the role of sediment in the transport of pollutants. Since much of the sediment transported in streams is of pedogenic origin, soil science has played a major role in determining the impact of soil properties on pollutant transport. As a soil scientist involved in the study of fluvial transport of soil (sediment), my attention is drawn to those soil properties which may influence the transport of pollutants irrespective of whether those pollutants were derived from land activities (eg. fertilizer additions to soil) or from non-rural sources (eg. point source discharges of heavy metal effluents). The objective of this paper, then, is to draw on our knowledge of soils to understand the role of sediment in the dynamics of pollutant transport in the fluvial environment, and to emphasize the chemically reactive nature of sediment, especially the clay-sized particles.

* Agronomy Department, The Ohio State University, Columbus, Ohio, USA

PHYSICAL-CHEMICAL PROPERTIES OF SOIL CLAYS

Soil is made up of a complex heterogeneous mix of primary and secondary minerals, organic compounds and exchangeable or labile chemical constituents. While the sand and silt components constitute a significant component of the total mass, it is the clay fraction that possesses the physico-chemical properties so important in pollutant transport. During the process of soil formation many of the products of weathering are concentrated in the clay fraction. These products include the secondary clay minerals such as illite, kaolinite, vermiculite and smectite, crystalline and amorphous iron, aluminum oxides and humified organic matter. The crystalline clay minerals possess a net negative permanent charge resulting from substitution of Mg^{2+} for Al^{3+}, and Al^{3+} for Si^{4+} in the aluminosilicate crystal structure during their formation. The iron and aluminum oxides, on the other hand, have a variable charge (net negative or positive) depending on the pH of the environment which determines the degree of protonation or de-protonation of the water molecules in coordination with the metal ions as shown in Figure 1.

Figure 1   Deprotonation of iron oxide with
            increasing pH

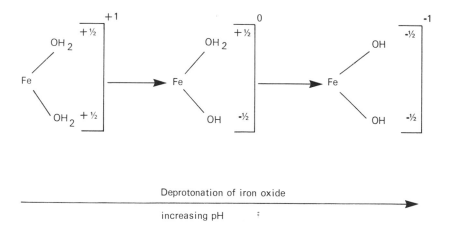

In soil systems, the cyrstalline aluminisilicates are often coated with amorphous or weakly crystalline hydrous oxides and so the particle may possess mixed characteristics.

The major properties of the clay-sized minerals of some significance in fluvial transport include:

1.   Large surface area and high adsorption capacity

Grim (1968) reports surface area measurements by BET adsorption of 15-100 m^2/g. This represents primarily external surface adsorption. The theoretical total surface area (internal and external surfaces) of montmorillonite (smectite) is 800 m^2/g. The crystalline aluminosilicate clay minerals have a high adsorption capacity for cations ranging from 3-15 milliequivalents/100 g clay for kaolinite to 150 milliequivalents/100 g for vermiculite and smectite (Grim 1968). Humified organic matter may have a cation adsorption capacity as high as 200 meq/100 g, the negative charge being due to hydrolysis of surface functional groups such as carboxyl, and as such are pH-dependent. The hydrous oxides have variable charge and, therefore, can adsorb both cations and anions, depending on pH. The pH of zero point of charge (ZPC), the point at which all negative and positive charges are neutralized, is quite high (>8.5) for the iron and aluminum hydrous oxides (Stumm & Morgan 1970), and at the pH of most streams these materials will still have considerable anion adsorption capacity. The hydrous oxides also have the capacity to specifically adsorb anions such as phosphate by ligand exchange (Figure 2).

Figure 2   Adsorbtion of phosphate anion by ligand exchange

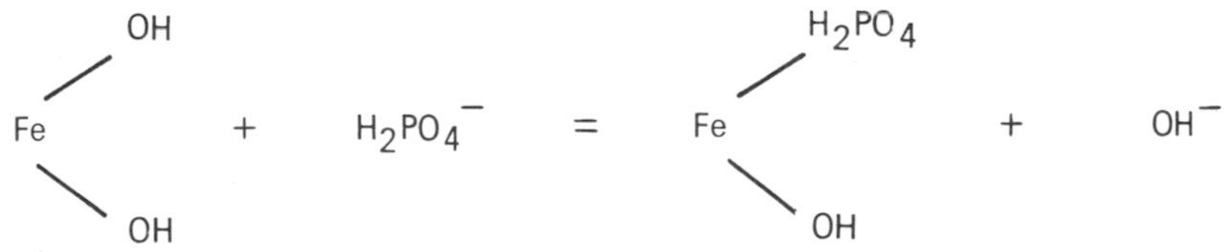

This capacity has been reported by Logan & McCallister (1977) to be as high as 4800 ug P/g sediment for stream bottom sediments in the Maumee River Basin of Ohio.

2.   Coagulation/dispersion

The high surface charge of the clay-sized sediment particles results in their mutual repulsion. This, together with Brownian motion and turbulence, serve to keep the particles relatively dispersed and, therefore, less susceptible to coagulation and settling. Adsorption of polyvalent cations such as Ca^{2+} and Mg^{2+} shield the negative charges on clay and allow them to approach close enough for coagulation and aggregate growth. Stumm & Morgan (1970) provide an excellent discussion of this process. Work by Wilding & Wall (unpublished) and Thompson et al (1976) in the Maumee River Basin demonstrate that much of the clay-sized sediment is aggregated into larger particle sizes because of the high levels of Ca^{2+} in this river which drains limestone glacial till and lacustrine sediments. Sediments suspended in septic tank effluent high in Na^+, a dispersing cation, exhibited far less tendency to coagulate. The common practice of determining particle size distribution of sediments by sedimentation following chemical dispersion may not reflect the particle size ranges actually found in the stream. Sedimentation of untreated suspended sediment following mild dispersion (eg. ultra sonic dispersion)

should be employed for stream sediments. Comparison with
the fully dispersed system would indicate the extent of
particle coagulation.

3.   Solute buffering capacity

Clay-sized soil and sediment regulate the concentration
of some dissolved chemical species because of the ability
of the particle to adsorb or desorb these materials.   Net
adsorption or desorption is determined by the concentration
of the adsorbate (solute) in solution and the equilibrium
concentration maintained by the adsorbent.   The adsorption
reaction can be described by a number of adsorption iso-
therms (Olsen & Watanabe 1957) which provide mathematical
relationships between solute adsorbed and the equilibrium
concentration of the solute.   The Langmuir isotherm
(Olsen & Watanabe 1957) for solutes adsorbed by soil or
sediment may take the form:

$$c/x/m = \frac{c}{b} + \frac{1}{kb}$$

c = equilibrium concentration of solute

x/m = weight of solute adsorbed per unit
weight of adsorbent (soil or
sediment)

k = constant related to strength of
adsorption bond

b = adsorption capacity of adsorbent

A plot of c/x/m versus c yields a linear curve with slope
$= \frac{1}{b}$ and ordinate intercept $= \frac{1}{kb}$. The constants derived
from this equation can be used to characterize adsorption
capacity and the strength with which the solute is held.

White & Beckett (1964) used a simple adsorption plot
to study the ability of colloids to buffer solute concentrations.   A plot of x/m versus c gives rise to the
relationship shown in Figure 3.

<u>Figure 3</u>   Plot of x/m versus c

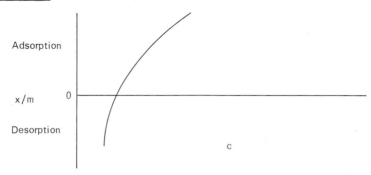

If the experiment is performed at low initial solute concentrations (the range of many dissolved species in streams)
there will be an initial desorption step which is dependent
on the amount of labile solute previously adsorbed.   As the
solute concentration increases, adsorption increases until

the adsorption capacity is exceeded.   Two useful parameters
can be extracted from this plot: c at x/m = o which
indicates the tendency of adsorbent to adsorb or desorb at
a given solute concentration (termed equilibrium phosphate
concentration, EPC).   The other parameter, the slope of the
curve at x/m = O is a measure of the buffer capacity of
the adsorbent, the amount of solute that must be adsorbed
or desorbed to effect a change in solute concentration.

Adsorption/desorption kinetics can be described by
the general equation:

$$x/m = KC_o t^{\frac{1}{n}}$$

C_o = Initial solute concentration

x/m = weight of solute adsorbed per unit weight of adsorbate

t = time K, n = constants

If log x/m is plotted against log t, a linear curve is obtained with the constants K and n derived from the slope
and intercept.   For preferentially adsorbed species such as
phosphate, adsorption is initially rapid and then continues
at a much slower rate over long periods (months).   Desorption
is similar to the slow adsorption process and is believed
to be diffusion controlled.   The implications of sediment
adsorption/desorption of phosphate will be discussed in a
later section.

CHEMICAL AND MINERALOGICAL TRANSFORMATION
OF SEDIMENT DURING FLUVIAL TRANSPORT

There have been attempts to use sediment mineralogy as an
indicator of sediment source.   Lund *et al* (1972) found that
the mineralogy of reservoir sediments in Illinois and
Indiana was similar to that of the watershed soils.   Klages
& Hsieth (1975) used silt and clay mineralogy to identify
sediment sources in the Gallatin River in Montana.   While
considerable variation occurred in the smaller tributaries
at different periods, reflecting a change in source area,
these differences were diminished in the main stretch as a
result of mixing.

The major assumption in the use of mineralogy as a
tracer for sediment source areas is that the mineralogy does
not change during fluvial transport.   There is some evidence,
however, that some change may occur and that these changes
reflect alteration of the sediment environment and the
extent to which sediment is deposited and resuspended during
transport to the mouth.   Our work on the Maumee River Basin
soils (Maumee River Basin Pilot Watershed Study, Semi-
annual Report 1976) shows that there is an increase in the
illite content of bottom sediments compared to the Basin
soils and an enhancement of the interlayer component of
expandable aluminosilicate clays.   The increase in illite
content may be due, in part, to selective transport of
illite since sediment leaving the watershed in runoff also
showed an increase in illite content.   Other changes in
sediment mineralogy noted were the formation of vivianite,
a ferrous iron phosphate and precipitation of secondary

calcite.   Calcite formation appears to be correlated with consumption of CO_2 by algal photosynthesis.   Frink (1969) studied lake sediments and found a large increase in illite content over watershed soils and a concomitant dechloritization of vermiculite.   Rolfe (1957) found that smectite dominated the clay fraction of upstream sediments while illite was predominant downstream.   These differences were attributed to differential deposition rather than diagenetic alteration.

On the other hand, Nabhan *et al* (1969) detected no differences in mineralogy of widely spaced sediments in the Nile River and Dewis *et al* (1972) found no differences in mineralogy of sediments from the fresh water drainage basin and the marine delta of the Mackenzie River, Canada. Moore (1961) found the clay mineralogy of Lake Michigan. bottom sediments to be similar to that of the glacial tills of the shore.

It would appear that when the bulk of suspended sediment is of surficial soil origin, the already weathered nature of these materials, especially the clay fraction, will preclude much additional diagenesis.

More sensitive than crystalline clay mineralogy in detecting changes in the sediment environment are changes in the chemistry of surface coatings, such as hydrous iron oxide.   Redox potential values as low as −300 mv have been recorded in the Maumee River Basin bottom sediments (Maumee River Basin Pilot Watershed Study, Semi-annual Report 1976) low enough to effect reduction of ferric iron.   Amorphous iron was higher in bottom sediments than surface soils in the Maumee River (Logan and McCallister 1977) and similar findings have been reported for lake sediments.

Sediment mineralogy has been used to distinguish between surficial and streambank erosion as a source of sediment (Wall & Wilding 1976),   In soils developed from calcareous parent material, weathering results in dissolution and leaching of carbonates to lower depths.   Therefore sediments derived from sheet erosion of surficial soil should be free of primary calcite, whereas streambank erosion of subsoil parent material should yield sediments containing. primary calcite.   Primary and secondary calcite may be distinguished by ^{14}C dating.

Another technique which may have some utility in distinguishing soil and streambank derived sediments is biogenic opal morphology.   Higher plant species are known to precipitate amorphous silica in intercellular spaces and these forms accumulate in the surface soil horizons after the plant has decayed.   Wilding & Drees (1971, 1973) were able to distinguish grass and tree sources of opal by their morphology, using scanning electron microscopy (SEM). These forms can also be distinguished from diatoms in the stream sediment.   Relative abundance of biogenic opal, then, would be indicative of the contribution of surficial soil to stream sediments.

SOIL AND SEDIMENT CHARACTERISTICS CONTROLLING THE TRANSPORT OF PHOSPHATE IN STREAMS

Concern during the last decade over the accelerated eutrophication of our lakes has prompted a wide range of studies on the factors affecting excessive growth of algae. The dominant role of phosphate (Hooper 1969) as a limiting nutrient in the growth of algae has been well documented. Phosphate is strongly adsorbed by soil and sediment by mechanisms discussed earlier. As a result, 90-95% of the total phosphate load entering lakes from streams is in the form of particulate P. Further, it is the P equilibrium characteristics of the sediment which determine the level of soluble P in the stream and lake, and the extent to which sediment P may be released to solution.

The total P content of soil is a function of the parent material geochemistry and ranges from 100 to 2000 ug/g with a mean of about 400 ug/g (Brady 1974). The addition of fertilizer or manure has little effect on the P content of the soil, although it does affect the release characteristics of soil phosphate. If we assume that plow-layer depth of an acre of soil weighs about 2×10^6 pounds, and a typical application of phosphate fertilizer to supply a corn crop is 50 pounds/acre, then the increase in total P content of the soil would be 25 ug/g. A 150 bushel/acre grain yield removes about 25 lb/ac or 12.5 ug/g (Ohio Agronomy Guide 1976), and so the increase in total P is 12.5 ug/g or 3% of the mean.

During the process of erosion and transport across the watershed, the finer clay is selectively transported; the clay contains a higher concentration of total P than the soil as a whole, and therefore the sediment will have a higher (enriched) P content than the soil from which it was derived. The enrichment ratio (ER), sediment P concentration/ soil P concentration, then can be used to calculate P loss by knowing soil loss. Porter (1975) has shown that ER decreases as soil loss increases, more of the silt and sand sized particles being eroded as total soil loss increases. ER is also a function of the particle size distribution of the soil, being much higher for sandy soils than for clay soils. An examination of the total P content of the clay fraction of Maumee River Basin watershed soils (McCallister 1976) indicated that particle size alone could not account for the observed increase in total P content of sediment in the stream (Logan & McCallister 1977). EPC values for Maumee River suspended sediments are higher than the average dissolved phosphate concentration in the stream. Therefore, considerable soluble phosphate must have been adsorbed by the sediment during erosion and initial transport.

Sediment can act as a source of P if the dissolved phosphorus concentration of the stream is lower than the sediment EPC. On the other hand, it can also serve as a sink for soluble P when stream P concentrations exceed EPC. Taylor & Kunishi (1971) showed a decrease in soluble P coming from a hog farm discharge into a small stream in Pennsylvania. The initial concentration was 2.23 ug/ml and

9.   Chemical and mineralogical indices

this was reduced by upstream sediment to 0.040 ug/ml downstream.   The sediment EPC was 0.010 ug/ml.   Much of the P
adsorbed by sediment is labile as evidenced by the higher
rate of P desorption by suspended sediment compared with
watershed soils in the Maumee (Green 1977).   Two points
are worth mentioning concerning P adsorption/desorption by
sediment: 1) The capacity to adsorb P is usually much
greater than the P actually adsorbed in the stream unless
the sediment is in contact with a point source such as
sewage effluent;   and the sediment appears to recover part
of its adsorption capacity with time, apparently as a
result of nucleation and crystal growth of the adsorbed P.
2) The rate of P desorption is much slower than the initial
rate of P adsorption.   Therefore the sediment responds more
rapidly as a sink than as a source.

SEDIMENT REACTIONS WITH HEAVY METAL AND PESTICIDES

The fine grained sediment also reacts with trace metals,
acting as a source for dissolved metals and as a sink when
it contacts high metal levels.   Naymik et al (1976) found
that the sediment concentration of chromium downstream
from an industrial chromium discharge was many fold higher
than upstream sediments.   Chemical extraction indicated that
the chromium adsorbed was labile and would be available for
uptake by stream fauna and flora.   Metals are held by
sediment by a number of mechanisms, adsorbed by the hydrous
oxides of Fe, Al and Mn, precipitated as metal carbonates
and complexed by the sediment organic fraction (Whitby
et al 1977).

Sediment is also the means by which most pesticides
are transported in the stream.   Long-lived compounds such
as DDT and its derivatives, aldrin, dieldrin and heptachlor
are concentrated in the sediment with soluble levels at or
below the level of detection.   Other chlorinated hydro-
carbons such as PCB's are also retained by the sediment.

SUMMARY

1.   The physical-chemical properties of suspended sediments
      reflect the chemical and mineralogical nature of the
      soils from which they are derived, and it is the clay
      fraction which most exhibits the reactive nature of
      sediment.
2.   Although there appears to be some evidence of mineral
      diagenesis during fluvial transport, the mineralogy of
      stream sediment generally reflects source area
      mineralogy.
3.   Phosphate is strongly adsorbed by clay-sized sediment,
      and the levels of soluble phosphate transported in the
      stream will be governed by the adsorption capacity and
      intensity of suspended sediment and the amount of
      labile phosphate on the sediment particle.
4.   A better understanding of sediment mineralogy and
      chemistry is needed in order to predict the impact
      of pollutants on in-stream water quality and the
      effects of tributary pollutant loads on estuary and
      lake quality.

REFERENCES CITED

Brady, N.C., 1974, *Nature and properties of soils.* Macmillan Co., New York, 639pp

Dewis, F.J., Levinson, A.A. & Bayliss, P, 1972, Hydrogeochemistry of the surface waters of the Mackenzie River drainage basin, Canada - IV. Boron-salinity-clay mineralogy relationships in modern deltas. *Geochemica et Cosmochimica Acta,* 36, 1359-1375

Frink, C.R., 1969, Chemical and mineralogical characteristics of eutrophic lake sediments. *Soil Science Society of America Proc.,* 33, 369-372

Green, D.B., 1977, *Calcite occurrence, stability, and phosphorus interactions in the fluvial media exiting the Maumee River drainage system.* MSc Thesis, The Ohio State University

Grim, R.E., 1968, *Clay mineralogy,* McGraw-Hill, New York, 596pp

Hooper, F.F., 1969, Eutrophication indices and their relation to other indices of ecosystem change. In: *Eutrophication: causes, consequences, correctives.* National Academy of Sciences, Washington DC, 225-235

Klages, M.G. & Hsieth, Y.P., 1975, Suspended solids carried by the Gallatin River of southwestern Montana: II. Using mineralogy for inferring sources. *J. Environmental Quality,* 4, 68-73

Logan, T.J. & McCallister, D.L., 1975, Phosphorus adsorption/ desorption relationships in Maumee River Basin soils and sediments. *20th Conference on Great Lakes Research, Abstracts.*

Lund, L.J., Kohnke, H. & Paulet, M., 1972, An interpretation of reservoir sedimentation II. Clay mineralogy. *J. Environmental Quality,* 1,303-307

Maumee River Basin pilot watershed study. Pollution from land use activities reference group. 1976. Semi-annual report, Ohio State University

McCallister, D.L., 1976, *Losses and adsorption-desorption characteristics of phosphorus from agricultural soils in the Maumee River Basin of Ohio.* MSc Thesis, The Ohio State University

Moore, J.E., 1961, Petrography of northeastern Lake Michigan bottom sediments. *J. Sedimentary Petrology,* 31, 402-436

Nabhan, H.M., Sys, C. & Stoops, G., 1969, Mineralogical study of the suspended matter in the Nile Water. *Pedologie,* 19, 30-38

Naymik, T.G., Wilding, L.P. & Logan, T.J., 1976, Variability of selected heavy metals, Maumee River Basin, NW Ohio, *19th Conference on Great Lakes Research, Abstracts,* 82

9.   Chemical and mineralogical indices

Ohio Agronomy Guide, 1976 Cooperative Extension Service. *The Ohio State University, Bull.*, 472

Olsen, S.R. & Watanabe, F.S., 1957, A method to determine a phosphorus adsorption isotherm of soils as measured by the Langmuir adsorption isotherm. *Soil Science Society of America Proc.*, 21, 144-149

Porter, K.S., 1975, *Nitrogen and phosphorus, food production, waste and the environment.* Ann Arbor Science, Ann Arbor, Michigan, 372pp

Rolfe, B.N., 1957, Surficial sediments in Lake Mead. *J. Sedimentary Petrology*, 27, 378-386

Stumm, W. & Morgan, J.J., 1970, *Aquatic chemistry.* Wiley Interscience, New York, 639 pp

Taylor, A.W. & Kunishi, H.M., 1971, Phosphate equilibria on stream sediment and soil in a watershed draining an agricultural region. *J. Agricultural Food Chemistry,* 19, 827-831

Thompson, M.L., Wilding, L.P., Smeck, N.E., & Vepraskas, M.J., 1976, Clay flocculation in fluvial media. *Agronomy Abstracts, 1976 Annual meeting, American Society of Agronomy*

Wall, G.J. & Wilding, L.P., 1976, Mineralogy and related parameters of fluvial suspended sediments in northwestern Ohio. *J. Environmental Quality,* 5, 168-173

Whitby, L.M., Skinner, D.S. & Strzelczyk, Z.S., 1977, The distribution of selected heavy metals in the sediments of agricultural watersheds. *20th Conference on Great Lakes Research, Abstracts*

White, R.E. & Beckett, P.H.T., 1964, Studies on the phosphate potential of soils: Part 1 - The measurement of phosphate potential. *Plant and soil,* 20, 1-6

Wilding, L.P. & Drees, L.R., 1971, Biogenic opal in Ohio soils. *Soil Science Society of America Proc.,* 35, 1004-1010

Wilding, L.P. & Drees, L.T., 1973, Scanning electron microscopy of opaque opaline froms isolated from forest soils in Ohio. *Soil Science Society of America Proc.,* 37, 647-651

RESUME

UNE MANIERE D'ABORDER LE PROBLEME DE LA PREDICTION DES COURBES D'EVALUATION POUR LE SEDIMENT

W. F. RANNIE

Les estimations de sediment en suspension pour la plupart des ruisseaux sont normalement obtenues des courbes représentant des taux de sédiment de la forme $G_s = kQ^j$ ou G_s = le débit de sédiment et Q = le débit d'eau. Dans cette étude on montre que la constante k et l'exposant j pour évaluer les courbes d'une grande variété d'environnements soumis à l'érosion sont liés par l'équation empirique j = 1.581 - 0.155 log k (: cfs - unités de tons par jour). Cette relation implique que les courbes d'évaluation peuvent être représentées par un réseau de lignes qui s'intersectent à un point commun et qui ont la forme générale $G_s = czQ$ 1.581 - 0.155 log c - 0.155 log z où k = cz et j^s= 1.581- 0.155 log k. On suppose que les paramètres c et z reflètent les caractéristiques physiques du bassin hydrographique qui déterminent le transport systématique d'alluvion.

Une regression multiple de $\log_k$ et une série de variables morphoclimatiques pour cinquante bassins hydrographiques répandus en divers endroits des Etats-Unis jusque dans le sud de l'Ontario ont donne les estimations les plus précises pour c et z soit:

$$c = 118.0; \quad z = q^{-2.304} \, H^{-1.209} \, A^{-0.676} \quad (r = 0.85)$$

$$c = 239.0; \quad z = q^{-2.381} \, x^{-1.511} \, A^{-0.677} \quad (r = 0.86)$$

où q = le moyen écoulement annual (cfs/mi^2), H = le relief maximum (pd.), X = le relief moyen (pd.) et A = l'étendue du bassin hydrographique (mi^2).

Pour les cinquante bassins sous étude les moyennes erreurs en prédisant le débit en suspension maximum étaient 143% et 147%. En examinant les restes il paraît que la variation inexpliquée est contribuée pour la plupart par les bassins d'enregistrement de courte durée et par les bassins dans la zone climatique méditerranée. On conclut que les fonctions générales d'évaluation proposées par l'auteur prédisent le transport de sédiment en suspension de longue durée pour les bassins avec des régimes climatiques continentaux avec du moins autant de précision que les autres méthodes disponibles et que ces fonctions d'évaluation n'ont besoin que du terrains etendus avec bassins qui sont facilement mesurés.

QUELQUES ASPECTS DE LA RESISTANCE FLUVIATILE DANS DES CHENAUX ESCARPES AVEC LITS GROSSIERS

T. J. DAY

L'étude des données provenant des expériences sur la résistance fluviatile pour les flots sous-critiques dans les chenaux avec lits grossiers revèle un départ radical des formules conventionnelles pour les valeurs basses de grossièrté relative. Une seule fonction de résistance n'existe pas pour les valeurs relatives de grossièrté de moins de 3, et des données limitées sont présentées qui indiquent que les caractéristiques de grossièrté dominent les caractéristiques fluviatiles. A présent on ne définit pas les relations cohérentes qui peuvent exister entre la résistance fluviatile, la configuration du lit fluvial et la déformation superficielle.

LA MECANISME ET LES ELEMENTS HYDRONAMIQUES DU TRANSPORT DE SEDIMENT DANS LES RUISSEAUX ALLUVIAUX
R. G. JACKSON II

L'analyse du transport de sédiment et de la hydraulique des chenaux ouverts peut fournir des informations substantielles sur la genèse et le fonctionnement des éléments fluviatiles qui concernent le géologue. Les sujets d'examen qui suivent exigent des techniques demandant des considérations multi-dimensionelles de la hydrodynamique et de la géométrie des chenaux et des considérations des caractéristiques détaillées des répartitions de sédiment alluvial de taille d'un grain: 1) séries des processus physiques du transport de sédiment (y inclus les structures fluviatiles et les lits fluviaux), 2) mécanismes et taux de transport de sédiment dans les ruisseaux et 3) répartition du matériel de taille de grain provenant du lit fluvial. Les tendances de la vitesse fluviatile, des lits fluviaux, de la géométrie des chenaux et de la grandeur du matériel provenant du lit fluvial d'un fleuve serpentant démontrent l'incapacité des méthodes à une dimension d'évaluer convenablement les trois sujets précédents. Pour expliquer de manière satisfaisante les éléments mentionnés ci-dessus aussi bien que d'autres éléments complexes des flots alluviaux on doit utiliser les méthodes sophistiquées de la représentation visuelle de flots et de particules, le mesurage des éléments directionnels des paramètres hydrauliques (surtout la vitesse), et la description suffisante des champs fluviatiles à quatre dimensions.

L'INTERPRETATION DE SUCCESSIONS ALLUVIALES ANCIENNES VU LES RECHERCHES MODERNES
B. R. RUST

On propose une nouvelle classification des réseaux de chenaux basée sur le paramètre anastomosé: le nombre d'anastomoses selon la longeur du méandre moyen. Les réseaux à un chenal et ceux à plusiers chenaux ont des paramètres anastomosés moins de et plus d'un respectivememt. Les réseaux sont subdivisés en petite et haute sinuosité à la limite de 1.5, qui donnent quatre types dont les chenaux uniformes de haute sinuosité (serpentants) et les chenaux multiformes de petite sinuosité (anastomosés) sont de loin les plus fréquents.

Le modèle pour les réseaux serpentants est bien établi: les séquences qui deviennent plus fines vers le haut avec une unité contenant un chenal qui est pour la plupart sablonneux, et avec une plaine d'inondation. Celui-là permet une estimation raisonnable des dimensions et des paramètres hydrologiques des paléochenaux.

Les dépôts anastomosés sont divisés en trois catégories soit: ceux qui sont dominés par le gravier, ou par le sable, ou par le silt; celui-ci est rare. Les graviers proximaux des hautes pentes (cones alluviaux) sont distingués de ceux des pentes plus basses par une variation lithologique plus grande, un raffinage vers le bas plus rapide, et (normalement) l'existence de dépôts provenant du débris fluviatile. Tous les graviers ont beaucoup d'imbrication stratifiée horizontalement déposée pour la plupart sur des bancs longitudinaux. Contraire aux opinions générales ces bancs sont considérés comme des lits d'équilibre étant stables sous les conditions d'inondations qui sont capables de déplacer tout le matériel du lit fluvial. Les graviers distaux sont caractérisés par des séquences qui deviennent plus fines vers le haut et qui sont dominées par des bassins de stratification entrecroisée à grande échelle contenant du gravier, mais ils possèdent aussi des unités fines formées sur des zones anastomosées inactives.

Les dépôts proximaux qui sont anastomosés et sablonneux démontrent une grande variabilité latérale et verticale avec beaucoup de surfaces érosives de "intraclasts" de pélite mais rarement des pélites primaires. Les dépôts distaux et sablonneux, au contraire, ont une continuité latérale et des séquences répétées et ils sont transitionnels dans des dépôts provenant de réseaux serpentants. Plusiers essais ont été faits d'estimer les dimensions et les paramètres hydrologiques des réseaux anciens anastomosés, mais à présent on est gêné par des connaissances insuffissantes des fleuves modernes anastomosés.

LES RECHERCHES INTERDISCIPLINAIRES SUR LES PROCESSUS D'ÉCOULEMENT ET D'ÉROSION À UN SITE ARIDE EXPÉRIMENTAL - DANS LE NORD DE NEGEV EN ISRAEL

A. YAIR

On décrit un programme de recherches adopté par un groupe interdisciplinaire qui étudie les processus d'écoulement et d'érosion à un bassin hydrographique expérimental étendant sur 11 325 milles. L'instrumentation du site consiste en appareils pour mesurer la précipitation, le vent, l'évaporation, l'humidité du sol, le taux et le rendement d'écoulement, le taux et le rendement de sédiment (minéral et dissous), la quantité et la répartition de sédiment biotique, la dynamique démographique des Isopodes (cloportes) et la contribution des particules de poussière. On analyse en détail la méthode et l'instrumentation employées pour l'étude de chacun des aspects variés contenus dans le programme de recherches. Les données obtenues jusqu'ici permettent une étude séparée de chacun des éléments mentionnés ci-dessus aussi bien qu'une analyse des relations complexes qui existent parmi eux. Le plan et le nombre d'instruments (plus de 100) permettent une etude precise des variations temporelles et spatiales qui ont lieu dans les processus et des elements qui les controlent. Finalement on indique quelques avantages qui proviennent de la méthode interdisciplinarie en ce qui concerne les recherches géomorphiques.

LES ELEMENTS HYDRAULIQUES QUI CONTROLENT LA MIGRATION DE CHENAUX
E. J. HICKIN

Les observations de la rivière Beatton indiquent que beaucoup de ses méandres ont été formés par des phases de migration discontinues. Chaque phase se termine quand un segment de chenal développe une courbure de proportion critique (rm 2.0), un phénomène préalablement expliqué par l'auteur en termes de la théorie Bagnold de séparation/ écoulement. Les observations sur les méandres notamment courbés de la rivière Beatton indiquent, cependant, que la séparation de la rive convexe ne s'y passe pas; la séparation fluviatile se passe au contraire aux rives concaves. Bien que la séparation de la rive concave ne soit pas souvent observées dans les études de la canalisation, son absence peut être une conséquence de la prétension générale mais fausse que le chenal est d'une largeur constante; on trouve quelques preuves de la séparation des rives concaves dans d'autres études fluviatiles. Ce phénomène fournit une explication hydraulique de la migration arrêtée des chenaux, ce qui est une autre solution possible à la théorie Bagnold de séparation/écoulement. On suggère des types spécifiques de recherches plus amples et on ébauche brièvement la possibilité d'une application pour l'ingénieur en ce qui concerne la séparation induite des rives concaves.

SEDIMENT EN SUSPENSION ET LES CARACTERISTIQUES DE LA REPONSE DE MATERIAUX DISSOUS
D. E. WALLING

Cet exposé note quelques résultats obtenus d'une étude de sédimentation fluviatile sur la rivière Exe et ses affluents et considère les implications plus larges. Un réseau de treize postes de mesurage a été établi dans le champ d'essai et on a utilisé 1) l'appareil pour l'enregistrement continu de conductance spécifique et des accumulations de sédiment en suspension et 2) l'équipment de prélèvement à la pompe automatique.

La variation d'accumulations totales de matériaux dissous à travers le champ d'essai pendant un temps de flot bas d'été a révélé une série de niveaux de conductance spécifique vingt fois plus grande. Le modèle associé a été contrôlé par le typé de roche et par l'utilisation de la terre. Les caractéristiques de matériaux dissous se

manifestant en temps de tempête démontrent un effet typique
de dilution quand on a égard aux accumulations totales de
solides dissous, mais une complexité beaucoup plus grande
est manifestée par la réaction d'ions individuels.

Les accumulations de sédiment en suspension montrent
aussi une variabilité spatiale bien prononcée à travers le
champ d'essai et il y a une série dix fois plus grande de
sédiments en suspension. Les relations entre les
accumulations de sédiment en suspension et le débit aux
postes de mesurage ne démontrent pas nettement que
l'accumulation dépend du débit. On note aussi les effets
de l'hystérésis, de la saison et de l'exhaustion sur ces
relations. En particulier la disponibilité de sédiment et
l'effet d'exhaustion associé avec celui-là exercent un
contrôle important sur la réponse sédimentaire pendant les
événements individuels, pendant une série d'hydrographes
et pendant une période prolongée.

L'influence d'agrégation et de déroutement de sédiment
et la réponse des matériaux dissous ont été examinées en
détail employant les données obtenues des postes de mesurage.

DES MODELES TEMPORELS ET SPATIAUX OBSERVES DANS LES PROCESSUS FLUVIATILES ET DANS L'EROSION
W. T. DICKINSON & G. J. WALL

On examine la variabilité temporelle et spatiale des processus
d'érosion et de sédimentation fluviatile pour les régions du
Canada à l'est des Rocheuses. Un modèle spatial a été
déterminé pour l'index d'érosion due à la précipitation ainsi
que pour les différences systématiques dans la répartition
mensuelle de l'index. Puisque le modèle saisonnier ne
coincide ni avec le modèle pour l'écoulement ni avec les
réponses du débit de sédiment en suspension, un index pour
l'érosion due à la précipitation printannière a été
déterminé et cartographié. Des diagrammes présentant le
régime de sédiment en suspension, les courbes de duration
sans dimension et l'analyse des valeurs extrèmes revèlent
le lieu et le temps de l'écoulement de sédiment fluviatile
et sa variabilité à travers le pays. Etant donné les
observations sur les modèles temporels et spatiaux, on émet
une hypothèse sur une perspective concernant l'érosion.

LES INDICES CHIMIQUES ET MINEROLOGIQUES
T. J. LOGAN

On discute la chémie et la minéralogie de sédiments en suspension
pendant le transport fluviatile comme indices de l'origine du
sédiment et comme indicatrices aussi de l'érosion surficielle et
littorale. On discute également les changements qui ont lieu
dans la chémie et la minéralogie du sédiment au course du transport fluviatile, et on décrit le rôle de sédiments dans le
transport de matériaux polluants, et dans la régulation des

formes solubles de matériaux polluants. On souligne l'importanc
du rôle de la chémie sédimentale et de la minéralogie dans la
coagulation de l'argile et on souligne l'adsorption/désorption
élémentaires dans l'environnement fluviatile.

0

DATE DUE